Methods in Molecular Biology™

Series Editor
John M. Walker
School of Life Sciences
University of Hertfordshire
Hatfield, Hertfordshire, AL10 9AB, UK

For further volumes:
http://www.springer.com/series/7651

Odontogenesis

Methods and Protocols

Edited by

Chrissa Kioussi

College of Pharmacy, Department of Pharmaceutical Sciences, Oregon State University, 1600 SW Jefferson Street, Corvallis, Oregon 97333, USA

Humana Press

Editor
Chrissa Kioussi, Ph.D.
College of Pharmacy
Department of Pharmaceutical Sciences
Oregon State University
1600 SW Jefferson Street
Corvallis, Oregon 97333
USA

ISSN 1064-3745 ISSN 1940-6029 (electronic)
ISBN 978-1-4939-5947-1 ISBN 978-1-61779-860-3 (eBook)
DOI 10.1007/978-1-61779-860-3
Springer New York Heidelberg Dordrecht London

Printed on acid-free paper

Humana Press is a brand of Springer
Springer is part of Springer Science+Business Media (www.springer.com)

Preface

Tooth repair has been around as long as human civilization. Only in recent times has the possibility of growing and implanting replacement teeth, made from one's own cells, moved into the realm of realistic possibilities. Recent advances show that stem cells can be induced from patient biopsies. However, growing teeth from such stem cells remains a formidable task that will require much detailed research into the molecular mechanisms of odontogenesis. Teeth are one of the most durable remains of vertebrates and, like bones, provide an interesting window on the evolution of body form when analyzed by comparative and molecular anatomy. The molecular and cellular mechanisms of tooth development must be studied in a range of vertebrates, from zebrafish to mice, so that evolutionarily conserved network kernels, which will define the cellular states of generic vertebrate tooth development, can be recognized.

Network kernels typically consist of five to ten recursively linked transcription factor genes that maintain each other's expression in particular cell types. They have begun to establish themselves as the predominant means to identify cell types during those early phases of development when specialized, mature genetic markers are not yet expressed. For example, combinations of transcription factors, rather than expression of specialized genes such as amelogenin, must be used to identify the cell types that interact during the early epithelial–mesenchymal transitions that create tooth buds from the epithelium and mesenchyme of the first branchial arch. Identifying cell types in a developmental process as "network states" or "combinatorial codes of transcription factors" creates the framework for studying the interactions between these cell types and ends the long and, in my opinion, rather futile search for "new" genes to mark each early cell type. Identifying the cellular "pieces" that operate in the "tooth-development" mechanism is an essential prerequisite to understanding and reverse-engineering that mechanism. It is for this reason that so many chapters of this book are dedicated to the detection of gene expression, both at the RNA and protein level.

Correlating gene expression patterns with morphological events in both time and space is one of the most powerful means to identify the molecular toolkit employed by a developmental process. The relative importance of each tool in the kit must be ascertained by perturbation assays. These are the essential follow-ups to expression assays. What happens to tooth development when we remove, add more, or alter a specific molecular component? In general, we tend to think that the larger the effect of perturbing a component, the more important the component. However, interpretation of gene perturbation data is tricky because individual molecular components are typically used more than once, either by different cell types at the same time or by the same cell type at different times, within a developmental mechanism. It is difficult to selectively remove a molecular component in only one cell type at only one stage in development. Perturbations that affect earlier stages of tooth bud development are likely to cause many secondary, tertiary, or further downstream effects and ultimately will produce the more obvious macroscopic, morphological tooth phenotypes. While such phenotypes are dramatic and may publish well, they will provide little functional information unless they are analyzed early enough, at time of first molecular divergence from wild type. Only perturbations that affect just the final stages of tooth development will produce effects that are correctly interpretable by looking at adult teeth. Several chapters of the book are dedicated to current approaches to the manipulation of gene expression levels and subsequent analysis of tooth phenotypes.

Last but not least, I have included a number of chapters concerning current efforts to get living tooth implants working without waiting for a full understanding of the developmental pathways at the molecular level. Such direct approaches often lead to working, practical solutions that help people well before the ideal solution is possible. Moreover, the methods being developed will ultimately be needed to get any prospective ideal "bioengineered tooth" into the patient's mouth.

Corvallis, OR, USA *Chrissa Kioussi*

Contents

Contributors

KENTARO AKIYAMA • *Center for Craniofacial Molecular Biology, University of Southern California, Los Angeles, CA, USA*
MARGRIT AMSTAD-JOSSI • *Department of Oral Biology, Center of Dental Medicine, University of Zurich, Zurich, Switzerland*
MALGORZATA M. BARCZYK • *Department of Biomedicine, University of Bergen, Bergen, Norway*
BRIAN K. BAY • *School of Mechanical, Industrial, and Manufacturing Engineering, Oregon State University, Corvallis, OR, USA*
ANNE ISINE BOLSTAD • *Department of Clinical Dentistry–Periodontics, University of Bergen, Bergen, Norway*
JAVIER CATÓN • *Departamento de Anatomía y Embriología Humana I, Faculty of Medicine, Complutense, University of Madrid, Madrid, Spain*
CHIDER CHEN • *Center for Craniofacial Molecular Biology, University of Southern California, Los Angeles, CA, USA*
NAOKI FUJIWARA • *Division of Developmental Biology & Regenerative Medicine, Department of Anatomy, Iwate Medical University, Iwate, Japan*
WILLIAM V. GIANNOBILE • *Michigan Center for Oral Health Research, School of Dentistry, University of Michigan, Ann Arbor, MI, USA*
STAN GRONTHOS • *Mesenchymal Stem Cell Group, Department of Haematology, Institute of Medical and Veterinary Science/Hanson Institute, Adelaide, SA, Australia*
DONALD GULLBERG • *Department of Biomedicine, University of Bergen, Bergen, Norway*
HIDEMITSU HARADA • *Division of Developmental Biology & Regenerative Medicine, Department of Anatomy, Iwate Medical University, Iwate, Japan*
MASAKI J. HONDA • *Department of Anatomy and Dental Research Center, Nihon University School of Dentistry, Tokyo, Japan*
ANN HUYSSEUNE • *Evolutionary Developmental Biology, Ghent University, Ghent, Belgium*
ANNE-MARTHE JEVNAKER • *Department of Oral Biology, University of Oslo, Oslo, Norway*
QIMING JIN • *Department of Periodontics and Oral Medicine, School of Dentistry, University of Michigan, Ann Arbor, MI, USA*
CHRISSA KIOUSSI • *College of Pharmacy, Department of Pharmaceutical Sciences, Oregon State University, Corvallis, OR, USA*
SERGIY KYRYACHENKO • *Department of Pharmaceutical Sciences, College of Pharmacy, Oregon State University, Corvallis, OR, USA*
KATERYNA KYRYLKOVA • *Department of Pharmaceutical Sciences, College of Pharmacy, Oregon State University, Corvallis, OR, USA*
MARIA A. LANDIN • *Department of Oral Biology, University of Oslo, Oslo, Norway*

MARK LEID • *Department of Pharmaceutical Sciences, College of Pharmacy, Oregon State University, Corvallis, OR, USA*
HANS U. LUDER • *Department of Oral Biology, Center of Dental Medicine, University of Zurich, Zurich, Switzerland*
JULIE T. MARCHESAN • *Department of Periodontics and Oral Medicine, School of Dentistry, University of Michigan, Ann Arbor, MI, USA*
THIMIOS A. MITSIADIS • *Department of Orofacial Development and Regeneration, Faculty of Medicine, Institute of Oral Biology, ZZM, University of Zurich, Zurich, Switzerland*
PETER R. MORGAN • *Oral Pathology, King's College London Dental Institute, London, UK*
MIGUEL PADIAL-MOLINA • *Department of Periodontics and Oral Medicine, School of Dentistry, University of Michigan, Ann Arbor, MI, USA*
MIHO OGAWA • *Tokyo University of Science, Chiba, Japan*
ATSUSHI OHAZAMA • *Department of Craniofacial Development, Dental Institute, King's College London, London, UK*
MASAMITSU OSHIMA • *Tokyo University of Science, Chiba, Japan*
HARALD OSMUNDSEN • *Department of Oral Biology, University of Oslo, Oslo, Norway*
KEISHI OTSU • *Division of Developmental Biology & Regenerative Medicine, Department of Anatomy, Iwate Medical University, Iwate, Japan*
HECTOR F. RIOS • *Department of Periodontics and Oral Medicine, School of Dentistry, University of Michigan, Ann Arbor, MI, USA*
ELLEN SANDERS • *Evolutionary Developmental Biology, Department of Biology, Ghent University, Ghent, Belgium; Department of Molecular Biology, Ghent University, Ghent, Belgium; Molecular Cell Biology Unit, Department for Molecular Biomedical Research, VIB, Ghent, Belgium*
SONGTAO SHI • *Center for Craniofacial Molecular Biology, University of Southern California, Los Angeles, CA, USA*
ANDREI D. TAUT • *Department of Periodontics and Oral Medicine, School of Dentistry, University of Michigan, Ann Arbor, MI, USA*
SHUHEI TSUCHIYA • *Department of Anatomy, Nihon University School of Dentistry, Tokyo, Japan; Department of Oral and Maxillofacial Surgery, Nagoya University Graduate School of Medicine, Nagoya, Japan*
TAKASHI TSUJI • *Research Institute for Science and Technology, Tokyo University of Science, Chiba, Japan*
BARBARA VERSTRAETEN • *Evolutionary Developmental Biology, Department of Biology, Ghent University, Ghent, Belgium*
MASATO YASUKAWA • *Tokyo University of Science, Chiba, Japan*

Chapter 1

Histological Analysis of the Embryonic and Adult Tooth

Atsushi Ohazama

Abstract

Histology is the study of the microscopic anatomy of tissues by examining a thin slice of the tissue under the microscope. Prior to slicing/sectioning, most tissues require some form of solidifying to allow thin sections to be cut. However, since the tooth is the hardest substance in the vertebrate body, it is one of the most difficult tissues to process for histology. This chapter describes the methods used for making histological sections of tooth from different embryonic stages through to adulthood.

Key words: Fixation, Decalcification, Dehydration, Clearing, Embedding, Staining, Enamel, Dentin, Pulp, Cementum, Alveolar bone, Periodontal ligament, Gingival tissue, Paraffin wax, Resin

1. Introduction

The tooth can be described as comprising two parts—the crown which is visibly present in the oral cavity and the root which is buried in the gingival and tooth-supportive tissue. Enamel, dentin, cementum, and dental pulp are the four major structures present within the vertebrate tooth. The dentin component is covered by enamel on the crown and cementum on root. Enamel is the hardest and most highly mineralized substance in the body. Although dentin is also a calcified tissue, it is less mineralized and less brittle than enamel. Enamel consists of a tightly packed mass of hydroxyapatite crystals in an organized pattern (enamel rods), whereas dentin shows dentinal tubules which radiate outward through the dentin from the pulp to the cementum. The dental pulp is located in the central part of a tooth and is surrounded by dentin—a complex organ composed of nerve axons, blood vessels, and connective tissues. Although the tooth is the hardest substance in the body, it develops from two soft tissues—epithelium and mesenchyme. These two tissues interact

Chrissa Kioussi (ed.), *Odontogenesis: Methods and Protocols*, Methods in Molecular Biology, vol. 887,
DOI 10.1007/978-1-61779-860-3_1,

with each other throughout development. The first morphological sign of tooth development is a narrow band of thickened epithelium on the developing jaw primodia. The thickened epithelium progressively takes the form of bud, cap, and bell configurations as differentiation proceeds. Subsequently, epithelial cells and mesenchymal cells (dental papilla) differentiate into enamel-producing ameloblasts and dentin-producing odontoblasts, respectively. Some parts of the dental papillae develop into dental pulp. Mineralization takes place during the late phase of the bell stage.

The periodontium is the tissue surrounding the tooth and functions as a supporting structure. It consists of the cementum, periodontal ligaments, alveolar bone, and gingiva. The cementum is a specialized bony substance covering the root dentin. The periodontal ligament is a connective tissue that attaches the cementum of a tooth to the alveolar bone, forming the alveolus around the teeth. In common with crown development, the periodontium also develops from soft tissue—dental follicles which give rise to cementoblasts for cementum formation, osteoblasts for alveolae bone formation, and fibroblasts for periodontal ligament formation. Development of the periodontium and the tooth root takes place after crown formation is almost completed. Most tooth specimens for histological analysis are likely to contain periodontal tissues. This combination of mineralized structures and surrounding tissues, thus, makes the tooth one of the most difficult tissues to process for histological examination. Needless to say, extra care is required when processing the tissue to make histological sections of tooth.

Histological analysis is the examination of a thin slice (section) of tissue under a microscope. The aim of tissue processing for histology is to embed the tissue in a solid medium to give it sufficient rigidity to enable thin sections to be cut. Paraffin wax is most frequently used for histological analysis as an embedding matrix, since paraffin wax is similar in density to most soft tissues.

Enamel, dentin, and alveolar bone all contain calcium deposits that will not be sectioned properly with paraffin embedding due to the difference in densities between the calcium and paraffin. To allow sectioning, this calcium must be removed (decalcification) prior to paraffin embedding. Tooth enamel is the hardest and most highly mineralized substance in the body (about 96% mineral, with water and organic material accounting for the other 4%). After decalcification, the enamel layer is identified as a blank space. The structure of dentin can also be affected during decalcification, since dentin contains approximately 65% mineral. For enamel analysis or precise dentin observation, un-decalcified sections have to be selected. The type of histological section chosen is, thus, dependent on the purpose or target of histology.

Paraffin sections are suited for most embryonic stages since tissue mineralization takes place at around birth. The timing of mineralization has to be considered when choosing the type of histological sections to be used, since this varies dependent on the tooth type and species to be studied (see Note 1). Frozen sections are best suited when histological analysis is required in a hurry or for special analytical techniques, such as immunohistochemistry or in situ hybridization.

2. Materials

2.1. Fixative

1. 10% Neutral buffered formalin (4% formaldehyde in phosphate-buffered saline, PBS).
2. 4% Paraformaldehyde (PFA; see Note 2).
3. Bouin's fixative (saturated aqueous picric acid 75 ml, glacial acetic acid 5 ml, and 40% formaldehyde 25 ml).
4. Zamboni's fixative (2% PFA and 15% saturated picric acid in 0.1 M phosphate buffer).
5. Carnoy's fixative (100% ethanol:chloroform:acetic acid = 6:3:1).
6. 95% Ethanol (see Note 3).
7. 100% Methanol (see Note 3).

2.2. Decalcifying Agents

1. 12.5% EDTA containing 2.5% PFA (see Note 4).
2. 10% EDTA (pH 7.4) (see Note 4).
3. 10% EDTA/Tris–HCl (pH 7.4) (see Note 4).
4. 10% EDTA/0.07% (w/v) glycerol (pH 7.4) (see Note 4).
5. Morse solution (10% sodium citrate and 22.5% formic acid (1)) (see Note 5).
6. 5% Formic acid.
7. 10% Nitric acid (see Note 6).
8. 10% HCl (see Note 6).
9. 5% Trichloroacetic acid.
10. Plank–Rychlo's solution (0.3 M aluminum chloride, 3% HCl, and 5% formic acid (2)).

2.3. Dehydrating Agents

1. Ethanol.
2. Methanol (methyl alcohol).
3. Isopropanol (2-propanol, isopropyl alcohol; see Note 7).
4. Butanol (buthyl alcohol).
5. Industrial methylated spirit (denatured alcohol; see Note 8).
6. Acetone.

2.4. Clearing Agents

1. Xylene (see Note 9).
2. 1,2,3,4-Tetrahydronaphthalene (see Note 9).
3. Citrus fruit oils (orange terpenes, Limonene reagent).
4. Toluene (see Note 10).

2.5. Embedding Materials

1. Paraffin wax (see Note 11).
2. Optimal cutting temperature (OCT) compound.
3. Epoxy resin (Epon812).
4. Acrylic resin (methyl methacrylate (MMA)).
5. Polyester resin.
6. Araldite resin.

2.6. Sectioning

1. Microtome.
2. Cryostat (for frozen section).

2.7. Slides

1. Aminosilane-coated slides (see Note 12).
2. Poly-L-lysine-coated slides (see Note 13).

2.8. Staining

1. Hematoxylin (see Note 14).
2. Acidic alcohol solution (1% HCl in 70% EtOH).
3. 0.25% EosinY in ethanol solution (0.5 g EosinY in 50 ml distilled water + 150 ml 80% ethanol + 1 ml glacial acetic acid).
4. 1% EosinY solution (0.5 g EosinY in 50 ml distilled water).
5. Alcian blue solution (see Note 15).
6. 0.5% Chlorantine fast red 5B.
7. 1% Phosphomolybdic acid (shake before use).

3. Methods

Specimens should be kept moist throughout processing, since drying produces artifacts. The optimal parameters for tissue processing can be empirically adjusted. Extreme care should be taken during solution changes to avoid damaging the embryos.

Tissues have to be trimmed as small as possible to aid the penetration of the preparatory solutions. Skin with hair has to be removed, since hair prevents penetration of the solution. Removing skin or making incisions in the skull also aids penetration when the specimen is large or matured.

A number of artifacts can result from improper fixation, poor dehydration, infiltration of embedding agents, and poor microtome sectioning.

3.1. Fixation

Fixation is the process used to preserve tissue from degradation and to maintain the tissue structure. Fixation should be carried out as soon as possible after removal of tissue. A variety of fixatives are available for use, depending on the type of tissue and features to be preserved, since there is no perfect fixative. All fixatives have their advantages and disadvantages; some are restrictive while others are for multipurpose. Generally, most fixatives preserve tissues by irreversibly cross-linking proteins.

1. Immersion
   Immersion or perfusion fixation has to be chosen depending on the tissue type.
   The volume of fixative used is important for the complete immersion of specimens. There should be a 10:1 ratio of fixative to tissue. The size and density (maturity) of the tissue must also be taken into consideration when calculating fixation time. Larger or more mature specimens take longer for the fixative to penetrate the deeper tissues. Agitation of the specimen in the fixative also enhances the fixation process. Increasing the temperature will also increase the speed of fixation, as long as the tissues are not cooked. In addition, high fixative concentrations may damage the tissues and produce artifacts.
2. Perfusion
   Perfusion fixation should be performed when infiltration of fixative is prevented by the presence of enamel, dentine, or alveolar bone. Perfusion fixation is also recommended when detailed observations of structures, such as cytoplasmic features, are needed. In these instances, the animal has to be transcardially perfused with physiological saline, followed by fixative, under deep anesthesia. The tissues are then dissected and immersed in the same fixative. Perfusion also allows a reduction in the time of fixation.

3.2. Decalcification (for Paraffin Sections)

There are inherent difficulties in the process of preparation of histological section of tooth due to extensive mineralization. Decalcification, therefore, is an essential process for making paraffin sections of calcified tissue. A variety of agents or techniques have been used to decalcify tissue and none of them work perfectly. Decalcification time varies depending on the type of agents, size of specimens, and degree (hardness) of mineralization. Decalcification should be performed with a stir bar (while the specimen is wire suspended in solution) or agitation at 4°C. The decalcification solution should be changed several times a week. The completion of decalcification is determined by transmission of soft X-ray microCT scan or cutting of unrequired parts of the specimens by blade.

3.3. Dehydration

The tissue must be supported in a hard matrix to allow sufficiently thin sections to be cut. Since most of the hard matrix is immiscible with water, tissues are dehydrated using a graded alcohol series of increasing concentrations. The vial in which the tissue is processed should be tightly capped to prevent evaporation of the alcohols. Tissues can be stored indefinitely in 70% ethanol without harm. Tissues that are insufficiently dehydrated prior to clearing and infiltration with paraffin wax are hard to section on the microtome, resulting in tearing and holes within the sections.

3.4. Clearing

Clearing is the transition step between dehydration and infiltration with the embedding medium. It should be miscible with both the dehydration agent and the embedding material. Clearing time varies depending on the types of clearing agent used and the size and density (maturity) of specimens. The time taken to clear is an important factor to consider as some clearing agents harden the specimen. The completion of clearing can be determined easily with clearing agents, such as xylene and 1,2,3,4-tetrahydronaphthalene, as the tissue becomes transparent.

3.5. Infiltration of Embedding Matrix and Embedding

After clearing, the specimens are infiltrated with the supporting substance in which the tissues are to be embedded. A vacuum can be applied to assist penetration of the embedding agent. During embedding, tissues are placed into molds (embedding boat) along with liquid embedding material that hardens. The specimens have to be embedded towards the bottom of the embedding boat with fresh embedding agent, since sectioning will be performed from the bottom of the boat upwards. The specimens also have to be oriented into the required plane for sectioning.

3.5.1. Paraffin Sections

Paraffin wax is most frequently used for histological analysis. Shrinkage occurs when tissues are transferred from dehydrating to clearing agents, and from the clearing agent to wax. Paraffin wax with a melting point of 60°C is most frequently used for histological analysis, although a variety of paraffin waxes with various melting points are commercially available.

Example (1). Mouse head embryonic day (E) 14 by xylene

4% PFA overnight at 4°C.

PBS wash, three times on a shaker at room temperature (RT).

30% EtOH 30 min to 1 h on a shaker at RT.

50% EtOH 30 min to 1 h on a shaker at RT.

70% EtOH overnight (O/N) at 4°C.

80% EtOH 30 min to 1 h on a shaker at RT.

90% EtOH 30 min to 1 h on a shaker at RT.

100% EtOH 30 min, three times on a shaker at RT.

Xylene 20 min, three times on a shaker at RT (until tissue becomes transparent).

Xylene/wax 20 min at 60°C.

Paraffin wax 30 min to 1 h × three times at 60°C.

Embedding.

Example (2). Mouse head (E) 17 by 1,2,3,4-tetrahydronaphthalene
4% PFA overnight at 4°C.
PBS wash, three times on a shaker at RT.
30% MeOH 1 h on a shaker at RT.
50% MeOH 1 h on a shaker at RT.
70% MeOH O/N at 4°C.
80% MeOH 1 h on a shaker at RT.
90% MeOH 1 h on a shaker at RT.
100% MeOH 1 h on a shaker at RT.
100% MeOH O/N at 4°C.
100% MeOH/isopropanol 1 h on a shaker at RT.
100% Isopropanol O/N at 4°C.
Tetrahydronaphthalene 1–3 h on a shaker at RT (until tissue becomes transparent).
Tetrahydronaphthalene 2 min at 60°C.
Tetrahydronaphthalene/wax 30 min at 60°C.
Paraffin wax 1 h, five times at 60°C.
Paraffin wax O/N at 60°C.
Paraffin wax 30 min at 60°C.
Embedding.

3.5.2. Frozen Sections

Frozen sectioning is a rapid method to make sections of embryonic or soft tissues and is a preferred technique for immunohistochemistry and in situ hybridization. Sectioning and slide mounting have to be performed in a cryostat. Dehydration is not required in this process. The morphological quality of the frozen sections is lower than that of paraffin sections.

Example
4% PFA overnight or 1–3 h at 4°C (see Note 16).
PBS wash, three times at RT or 4°C.
10% Sucrose until specimens sink at 4°C (see Note 17).
20% Sucrose until specimens sink at 4°C (see Note 17).
30% Sucrose until specimens sink at 4°C (see Note 17).
Embedding (see Notes 18 and 19).
Store at −70°C.

3.5.3. Plastic Sections

Plastic embedding is required when thin sections (<4 μm thick) needed to be generated, since paraffin wax is not sufficiently hard for cutting very thin section. Enamel is visualized as a blank space in paraffin sections, since decalcification removes the enamel. Plastic embedding permits un-decalcified sectioning to be performed to allow observation of the enamel structure as long as the size of enamel is small in specimen.

Tissue processing is almost identical to that for paraffin embedding, up until dehydration. However, the time taken for each step should be longer due to the presence of calcified tissues. Processing should be performed according to the manufacturer's protocol, since each commercially available resin reacts differently (e.g., polymerizing by heat, ultraviolet light, chemical catalysts, etc.). Most plastic embedding compounds are hazardous and require fume control facilities and careful handling.

3.5.4. Ground Sections (Un-decalcified Sections)

The ground sectioning technique is used to make un-decalcified sections of specimens that additionally contain metal material. Ground sections are also suitable for enamel or dentin observation, including mineralization analysis by tetracycline labeling. Specimens are processed into plastic embedding materials and thick sections generated (150–300 μm). These thick sections are then polished to a thickness of around 40 μm followed by the application of alumina-polishing paste. Tissue processing is identical to that for paraffin embedding, up until dehydration. However, the time taken for each step should be longer due to the presence of calcified tissues. Processing after the dehydration and cutting/grinding steps should be performed according to the manufacturer's protocol (3). Most ground sectioning compounds are hazardous and require fume control facilities and careful handling.

Example. Postnatal mouse jaw labeled by tetracycline

4% PFA overnight at 4°C.
PBS wash, three times on a shaker at RT.
70% EtOH 30 min to 1 h on a shaker at RT.
80% EtOH 30 min to 1 h on a shaker at RT.
90% EtOH 30 min to 1 h on a shaker at RT.
95% EtOH 30 min to 1 h on a shaker at RT.
90% EtOH O/N at 4°C.
100% EtOH 30 min–1 h, two times on a shaker at RT.
Acetone, 100% EtOH (50:50, v/v) 30 min to 1 h, three times on a shaker at RT.
Acetone, two times on a shaker at RT (until tissue becomes transparent).
Acetone:plastic embedding solution (50:50, v/v) 4–5 h at RT.
Acetone:plastic embedding materials (25:75, v/v) O/N at RT.
100% plastic embedding materials 1 h × three times O/N at RT.
100% plastic embedding materials at 40°C (curing units) for polymerization O/N.
Change the temperature of curing units to 50°C for polymerization O/N.
Change the temperature of curing units to 60°C to complete polymerization O/N.

3.6. Sectioning

Once the block containing the tissues has hardened, it is ready for sectioning. A very sharp knife is critical for proper sectioning. The embedded block is mounted onto sectioning stubs, trimmed, and sectioned with a microtome, following standard procedures. Paraffin wax blocks should be kept at 4°C before sectioning, to retain the hardened composition. Sections are floated on warm distilled water, which helps to remove wrinkles. Sections are placed onto slides and slides are dried on a warmed hot plate (40–42°C) O/N to improve section–slide adhesion.

Frozen sectioning is performed in a cryostat. The temperature inside the cryostat is about −20 to −30°C. Sections are cut and picked up onto a slide and ready for use.

Common artifacts observed in the sections are tearing, ripping, holes, folding, wrinkles, artificial spaces, streaks, missing parts, etc. These can and must be avoided during sectioning. When multiple examinations are required of one specimen (e.g., histology, in situ hybridization, and immunohistochemistry), serial sections are made and divided over several slides, with each "set" used for a different analysis (see Note 20).

3.7. Staining

Most staining cannot be performed properly on tissue containing paraffin or plastic. Therefore, sections have to be deplastified or dewaxed by clearing agents followed by rehydration prior to staining.

Hematoxylin and eosin (H&E) staining is the most widely used staining method. Hematoxylin has an affinity for nucleic acids of the cell nucleus, whereas eosin has an affinity for cytoplasmic components of the cells. The variety of hematoxylin solutions available for use is based partially on the choice of metal ion used, which varies in intensity or hue (see Note 14). There are two ways to stain with hematoxylin depending on the type used. Slides can be left in the hematoxylin solution for a set period of time and then placed into a solution, such as acid–alcohol, to partially remove the stain. Alternatively, the slides are dipped into the hematoxylin solution until the desired intensity of staining is achieved. There are also a variety of eosins available for use and selection is dependent on the hue required. Decalcified tissues can sometimes become over-stained by eosin.

Alcian blue–chlorantine fast red staining additionally highlights the developing dentin and enamel, as well as bone and cartilage, through the application of a broad range of staining colors.

Slides bearing the stained sections must be covered with a thin piece of plastic or glass coverslip to protect the tissue from being scratched, to provide a better optical quality for viewing under the microscope, and to preserve the tissue sections over time. Mounting media should be used as an adhesive between the slide and coverslip. Xylene-based mounting media is the most widely used mounting reagent, although many other types including aqueous mounting media are available. For xylene-based mounting reagents, the stained slides are taken through a series of alcohol solutions to remove water, and then through clearing agents to a point at which a permanent resinous substance can be placed over the sections, beneath the coverslip.

3.7.1. H&E Staining

Example

All steps are performed at RT.
Xylene 5–10 min.
Xylene 5–10 min.
100% EtOH 5 min.
100% EtOH 5 min.
95% EtOH 5 min.
90% EtOH 5 min.
80% EtOH 5 min.
70% EtOH 5 min.
50% EtOH 5 min.
30% EtOH 5 min.
Tap water 5 min.
Distilled water briefly.
Gill's hematoxylin 5 min.
Running tap water 10 min.
Acidic alcohol solution 5–10 s (see Note 21).
Tap water 5 min.
Distilled water briefly.
Eosin solution 10–20 s.
70% EtOH briefly.
80% EtOH briefly.
90% EtOH briefly.
95% EtOH briefly.
100% EtOH 1 min.
100% EtOH 3 min.
Xylene 5 min.
Xylene 5 min.
Coverslip with mounting medium.
Air dry in the hood overnight.

3.7.2. Alcian Blue–Chlorantine Fast Red Staining

Example

Xylene 5–10 min.
Xylene 5–10 min.
100% EtOH 5 min.
100% EtOH 5 min.
95% EtOH 5 min.
90% EtOH 5 min.
80% EtOH 5 min.
70% EtOH 5 min.
50% EtOH 5 min.
30% EtOH 5 min.
Tap water 5 min.
Distilled water briefly.
Ehrlich's hematoxylin 15 min.
Running tap water 10 min.
Distilled water briefly.
Alcian blue solution 10 min.

Distilled water 10 s.
1% Phosphomolybdic acid 10 min.
Rinse in distilled water 10 s.
Chlorontine fast red 10 min.
Rinse in distilled water 10 s.
Rinse in distilled water 10 s.
70% EtOH briefly.
80% EtOH briefly.
90% EtOH briefly.
95% EtOH briefly.
100% EtOH 1 min.
100% EtOH 3 min.
Xylene 5 min.
Xylene 5 min.
Coverslip with mounting medium.
Air dry in the hood O/N.

4. Notes

1. Mineralization starts at embryonic month 4 in deciduous dentition, whereas it starts from birth to 3 years old in permanent dentition (except the third molar) in humans. In mice, calcification takes place from E18.5.
2. Stir solution at 60–65°C to dissolve the PFA powder. Cool solution and store at −20°C to avoid oxidization by atmospheric oxygen to formic acid. Instead, keep the solution in a cool and dark place after preparation and use it as soon as possible.
3. Alcohol-based fixatives generally do not give good morphology but they are useful in specific cases, such as BrdU staining and immunohistochemistry.
4. EDTA is generally preferable for in situ hybridization and immunohistochemistry. EDTA-based solutions, however, need a long duration to complete decalcification in comparison with other solutions.
5. Morse solution-treated specimens can be used for in situ hybridization and immunohistochemistry (4).
6. 10% Nitric acid and 10% HCl solutions have to be neutralized by sodium sulfate before dehydration.
7. Isopropanol does not cause over-hardening or shrinkage of the tissues.
8. Industrial methylated spirit has the same physical properties as ethanol.

9. Tissues are rendered transparent and facilitate clearing endpoint determination when tissues are soaked in xylene or 1,2,3,4-tetrahydronaphthalene.
10. Toluene is more tolerant of small amounts of water left in the tissue in comparison with xylene.
11. Paraffin can be purchased that differs in melting point and hardness.
12. Place glass slides in stainless rack→rinse in distilled water→rinse in alcohol solution (acetic acid 400 μl in 95% ethanol 400 ml) for 3 min→air dry→soak in 3-aminopropyl-triethoxy saline (8 ml) in acetone (400 ml) for 1 min→rinse in acetone × two times→rinse in distilled water→air dry.
13. Place glass slides in stainless rack→rinse in distilled water→soak in 2 N NaOH solution for 1–2 h→rinse in distilled water × five times→soak in 0.01% poly-L-lysin solution for 1 h→rinse in distilled water→rinse in 100% ethanol for 3 min→air dry.
14. There are many different hematoxylins, such as Mayer's, Carazzi's, Lillie-Mayer's, Delafield's, Harris's, Ehrlich's, etc.
15. Filter 1% Alcian blue 8 G 50 ml + 1% acetic acid 50 ml + thymol 10 mg.
16. Fixation can be performed after sectioning.
17. Specimens are highly likely to float in the solution and then gradually sink.
18. Placing a little embedding matrix into the embedding boat with the specimen prevents the specimens from floating. Add more embedding matrix to cover the specimens ensuring that the specimen remains at the bottom.
19. Let the bottom of the mold touch the liquid nitrogen first and then place approximately the bottom-third of the mold into the liquid nitrogen.
20. If the sections are intended for in situ hybridization, the entire protocol should be carried out very carefully, as all steps and solutions must be RNase free.
21. The requirement of acidic alcohol depends on the type of hematoxyline used.

Acknowledgments

I thank Dr. James Blackburn and Dr. Masato S. Ota for critically reading the manuscript.

References

1. Morse M (1945) Formic acid-sodium citrate decalcification and butyl alcohol dehydration of teeth and bones for sectioning in paraffin. *J Dent Res* 24; 143–153.
2. Plank and Rychlo (1952) Ein schnellentkalkungsmethode. *Zbl Pathol* 89; 252–254.
3. Donath and Breuner (1982) A method for the study of undecalcified bones and teeth with attached soft tissues. The Säge-Schliff (sawing and grinding) technique. *J Oral Pathol* 11; 318–326.
4. Shibata Y., Fujita S., Takahashi H., Yamaguchi A. Koji T (2000) Assessment of decalcifying protocols for detection of specific RNA by non-radioactive in situ hybridization in calcified tissues. *Histochem Cell Biol* 113; 153–159.

Chapter 2

Determination of Gene Expression Patterns by Whole-Mount In Situ Hybridization

Sergiy Kyryachenko, Kateryna Kyrylkova, Mark Leid, and Chrissa Kioussi

Abstract

Whole-mount in situ hybridization (WISH) is a reliable and specific method to study three-dimensional patterns of gene expression. A labeled nucleic acid probe anneals to a complementary target sequence and is visualized and localized in the embryo. This chapter describes a sensitive method for WISH on mouse embryos using digoxigenin-labeled RNA probes. The technique can be used for the analysis of gene expression patterns during early stages of odontogenesis and in tooth explants.

Key words: Whole-mount in situ hybridization, Mouse embryos, Digoxigenin-labeled probe, Alkaline phosphatase–antibody conjugate

1. Introduction

Whole-mount in situ hybridization (WISH) is a very powerful and widely used tool in developmental biology because it allows the analysis of complex expression patterns of genes in whole embryos. The use of WISH is very helpful to provide a three-dimensional overview of gene expression pattern, since the three-dimensional reconstruction from serial sections can be cumbersome (1, 2). Nowadays, the method can be used to study expression of more than one gene in the same embryo, allowing spatial and temporal overlaps to be examined (3).

The method uses RNA probes—molecules of RNA—which are complimentary to the endogenous mRNA and have been synthesized with a particular label. These probes are hybridized to the target RNA of an embryo, and the label can be visualized by different detecting systems (4, 5). Since the discovery of in situ hybridization,

Chrissa Kioussi (ed.), *Odontogenesis: Methods and Protocols*, Methods in Molecular Biology, vol. 887,
DOI 10.1007/978-1-61779-860-3_2, © Springer Science+Business Media, LLC 2012

different labels and detection systems became available. For example, radio-, fluorescent-, and antigen-labeled bases are detected by autoradiography, fluorescence microscopy, or immunohistochemistry, respectively (2, 6, 7). In the protocol described here, probes are labeled with digoxigenin-11-uridine-5′-triphosphate (DIG-11-UTP) and visualized by an anti-digoxigenin conjugated to alkaline phosphatase (AP), which catalyzes a chromogenic reaction (8).

The method has been adapted to different types of tissues and/or embryos. Gene expression analysis by in situ hybridization can be also used for organ cultures to study the functions of the soluble regulatory molecules, which can be added directly to the culture medium or used to generate growth factor-soaked beads. It is often informative to section whole-mount embryos or tooth explants after labeling (9, 10). Hence, this is a very efficient method to study different stages of odontogenesis. However, it is more preferable to perform in situ hybridization on embryo sections for later stages of mouse development because of probe penetration issue into the whole embryo (9).

In this chapter, we present a technique that has proven effective in investigating gene expression pattern during early stages of mouse tooth development.

2. Materials

Prepare all solutions using ultrapure deionized water and analytical grade reagents. Store all reagents according to manufacturer's instructions or, if not applicable, at room temperature (unless indicated otherwise).

2.1. Embryo Preparation

1. 10-cm Petri dish.
2. Dissection microscope.
3. Dissection tools: Watchmaker's forceps.
4. Shaker.
5. 1.5-ml microfuge tubes.
6. Plastic transfer pipettes.
7. Phosphate-buffered saline (PBS): 137 mM NaCl, 2.7 mM KCl, 10 mM Na_2HPO_4, and 1.76 mM KH_2PO_4 in dH_2O, pH 7.4 (with HCl).
8. 4% Paraformaldehyde in PBS (see Note 1).
9. PBST: PBS containing 0.1% Tween 20.
10. Methanol.
11. 25, 50, and 75% methanol in PBST.

2.2. In Situ Hybridization

1. Rocker.
2. Hybridization oven.
3. 2-ml conical-bottom, O-ring, screw-cap tubes.
4. 5-ml polystyrene, round-bottom tubes.
5. DEPC-treated dH_2O (see Note 2).
6. PBST.
7. 25, 50, and 75% methanol in PBST.
8. 6% Hydrogen peroxide in PBST.
9. Proteinase K, 10 mg/ml stock in 100 mM Tris–HCl (pH 7.5), 10 mM EDTA; store in aliquots at −20°C.
10. Glycine.
11. 0.2% Glutaraldehyde/4% paraformaldehyde in PBST; store in aliquots at −20°C.
12. SSC: 20× stock: 3 M NaCl, 0.3 M sodium citrate in DEPC-treated dH_2O, pH 4.5.
13. DIG-labeled probe.
14. SDS, 10% stock.
15. Hybridization buffer (make fresh before use): 50% formamide, 5× SSC, 1% SDS, 50 μg/ml yeast tRNA, 50 μg/ml heparin in DEPC-treated dH_2O (see Note 3).

2.3. Posthybridization Washes and Preabsorption of Antibody

1. Rocker.
2. Hybridization oven.
3. Shaker.
4. Microcentrifuge.
5. Vortex mixer.
6. 2-ml conical-bottom, O-ring, screw-cap tubes.
7. 5-ml polystyrene, round-bottom tubes.
8. 1.5-ml microfuge tubes.
9. SSC, 20× stock.
10. SDS, 10% stock.
11. Solution 1 (Sol 1): 50% formamide, 4× SSC (pH 4.5), and 1% SDS in dH_2O (see Note 3).
12. 5 M NaCl.
13. 1 M Tris–HCl, pH 8.0.
14. 25% Tween 20.
15. Solution 2 (Sol 2): 0.5 M NaCl, 10 mM Tris–HCl (pH 8.0), and 0.1% Tween 20 in dH_2O (see Note 3).
16. RNase A.

17. Solution 3 (Sol 3): 50% formamide, 2× SSC (pH 4.5), and 0.2% SDS in dH_2O (see Notes 3 and 4).
18. 1 M maleic acid, pH 7.5.
19. MBST: 100 mM maleic acid, 150 mM NaCl, and 0.1% Tween 20. Adjust pH to 7.5 with solid NaOH.
20. Sheep serum (HISS): Heat inactivate at 60°C for 30 min; store in aliquots at −20°C.
21. Blocking buffer: 2% blocking reagent (Roche) in MBST; store in aliquots at −20°C.
22. Anti-digoxigenin-AP, fab fragments (Roche).
23. Embryo powder: (1) Homogenize E11.5–E13.5 mouse embryos in a minimum volume of ice-cold PBS; (2) add four volumes of ice-cold acetone, mix, and incubate on ice for 30 min; (3) collect the precipitate by centrifugation at 10,000 × *g* for 10 min, remove, and discard the supernatant; (4) wash the pellet with ice-cold acetone for 10 min and then spin again; (5) transfer the pellet to a clean piece of filter paper, spread, and allow air drying overnight; (6) grind it into a fine powder using mortar and pestle. Store desiccated at 4°C (see Note 5).
24. Antibody solution: Place 3 mg of embryo powder into a 1.5-ml microfuge tube. Add 0.5 ml of blocking buffer. Rotate tube in hybridization oven at 70°C for 30 min. Vortex for 10 min and cool on ice. Add 5 μl of HISS and 1 μl of anti-DIG AP. Shake gently on shaker at 4°C for 1 h. Spin in microcentrifuge at 4°C for 10 min. Take ~350 μl of the supernatant and dilute to 4 ml with 1% HISS in blocking buffer. Store at −20°C.

2.4. Post-antibody Washes and Staining

1. Rocker.
2. 2-ml conical-bottom, O-ring, screw-cap tubes.
3. 5-ml polystyrene, round-bottom tubes.
4. 24-Well plates.
5. MBST.
6. 2 M levamisole; store in aliquots at −20°C.
7. 1 M $MgCl_2$.
8. 2 M Tris–HCl, pH 9.5.
9. NTTML: 0.15 M NaCl, 0.1 M Tris–HCl (pH 9.5), 0.1% Tween 20, 50 mM $MgCl_2$, and 2 mM levamisole in dH_2O (see Note 6).
10. BM Purple AP substrate (Roche).
11. PBST, pH 4.5 (pH with phosphoric acid).

12. 0.1% Glutaraldehyde/4% paraformaldehyde in PBS; store in aliquots at −20°C.
13. PBST.
14. Glycerol.
15. 50, 75, and 85% glycerol in PBST.

3. Methods

Carry out all procedures at room temperature unless otherwise specified.

3.1. Embryo Dissection

1. Dissect embryos and place them in ice cold PBS.
2. After all embryos are dissected, transfer them into 1.5-ml microfuge tubes using a plastic transfer pipette (cut the tip to enlarge the opening).
3. Fix in 4% paraformaldehyde in PBS with gentle shaking at 4°C overnight.
4. Wash embryos 2× with PBST, 10 min each time.
5. Wash embryos 1× with 25, 50, and 75% methanol in PBST, 5 min each time.
6. Wash embryos 2× with 100% methanol, 10 min each time. You can store embryos in 100% methanol at −20°C for several months.

3.2. In Situ RNA Hybridization

1. Rehydrate embryos 1× in 75, 50, and 25% methanol in PBST, and 2× in PBST, 5 min each time.
2. Transfer embryos into 5-ml polystyrene, round-bottom tubes (see Note 7).
3. Bleach embryos with 6% hydrogen peroxide in PBST for 1 h on rocker.
4. Wash 3× with PBST, 5 min each time.
5. Treat embryos with 10 μg/ml of proteinase K in PBST (see Note 8).
6. Rinse with fresh-made, filtered, 2 mg/ml glycine in PBST for 30 s.
7. Wash 2× with PBST, 5 min each time.
8. Refix the embryos with 0.2% glutaraldehyde/4% paraformaldehyde in PBST for 20 min.
9. Wash 2× with PBST, 5 min each time.

10. Transfer the embryos into 2-ml conical-bottom, O-ring, screw-cap tubes, add 1 ml of hybridization buffer, and pre-hybridize for 1 h at 70°C, rotating in hybridization oven.
11. Remove hybridization buffer and add 0.5–1 ml of hybridization buffer containing the 4 μg/ml of a probe. Hybridize in hybridization oven at 70°C overnight (see Notes 9 and 10).

3.3. Posthybridization Washes and Preabsorption of Antibody

1. Remove probe from tubes (see Note 11).
2. Add 1 ml of Sol 1 to tubes and pour the solution with embryos into a 5-ml polystyrene, round-bottom tube. Rinse embryos 1× with Sol 1.
3. Wash 2× with pre-warmed Sol 1 at 70°C, 30 min each time.
4. Wash 1× with pre-warmed mix of Sol 1:Sol 2 (1:1) at 70°C for 10 min.
5. Wash 3× with Sol 2, 5 min each time.
6. Incubate the embryos with 100 μg/ml RNase A in Sol 2 at 37°C for 1 h.
7. Wash 1× with Sol 2 and 1× with Sol 3, 5 min each time.
8. Wash 2× with pre-warmed Sol 3 at 65°C, 30 min each time.
9. Wash 3× with MBST, 5 min each time.
10. Pre-block embryos with 1–1.5 ml of 10% HISS in the blocking buffer for 3–4 h. Put the tubes upright on a shaker.
11. Remove blocking solution from embryos, transfer embryos into 2-ml conical-bottom, O-ring, screw-cap tubes, and add 1 ml of the antibody solution. Incubate on rocker at 4°C overnight (see Note 12).

3.4. Post-antibody Washes and Staining

1. Transfer embryos into a 5-ml polystyrene, round-bottom tube. Wash 3× with MBST, 5 min each time.
2. Wash 8× with MBST, 1 h each time. Leave in MBST at 4°C overnight (see Note 13).
3. Wash embryos 3× with NTMTL, 5 min each time.
4. Transfer embryos into 2-ml conical-bottom, O-ring, screw-cap tubes. Replace NTMTL with 1 ml of BM purple, wrap tubes in foil, and incubate on a rocker for ~1 h–2 days.
5. Once a signal is at the desired intensity, wash 3× with PBST, pH 4.5, 5 min each time, keeping in dark.
6. Fix with 0.1% glutaraldehyde/4% paraformaldehyde in PBS at 4°C for 1 h–overnight.
7. Transfer embryos into 24-well plates.

8. Clear embryos in 50, 75, and 85% glycerol in PBS series. The embryos should sink in each solution. Store embryos in 85% glycerol in PBS (see Note 14).

4. Notes

1. Make fresh 4% paraformaldehyde each time. Preparation should be carried out inside a fume hood. Store it at 4°C for up to 1 week.
2. RNA is easily destroyed by ribonucleases that are extremely stable enzymes and can be present in untreated solutions. Hence, it is essential to avoid contamination of the RNA probe and solutions by RNase before RNA hybridization. It is highly advisable to use DEPC-treated dH_2O to inactivate RNase contamination. Add 1 ml of DEPC in 1 l of dH_2O or reagent, shake vigorously, leave overnight at room temperature, and autoclave.
3. It is recommended to make fresh hybridization buffer, Solution 1, Solution 2, and Solution 3 each time.
4. Adding of SDS into Sol 3 is optional.
5. It is always important to use embryo powder prepared from the species that you are studying.
6. It is advisable to add $MgCl_2$ and fresh aliquot of levamisole directly before use.
7. Up to four E10.5 embryos can be used per tube.
8. Proteinase K solution should be freshly made each time. Timing is important for treatment with proteinase K and should be adjusted based on the tissue of interest (for example, E7.0 for 2 min; E7.5 for 3 min; E8.5 for 4 min; E9.5 for 15 min; E10.5 for 23 min). Let the tube sit on the table on its side and roll it gently every ~3 min. Do not shake. Embryos are very fragile after Proteinase K digestion before the second fixation.
9. Quantities of the probe needed for the hybridization should be optimized for each probe.
10. Make sure that the volume of the hybridization buffer covers the embryos.
11. You can store the probes for recycle and use them once or twice.
12. You can store the antibody solution for recycle and use them once or twice.

13. You can repeat step 2. The larger the embryos, the longer the washes should be. The embryos can be left washing at 4°C without apparent loss of signal for up to 3 days.
14. Keep embryos in 85% glycerol in PBS for a few days before taking pictures.

References

1. Rosen B. and Beddington R.S.P. (1993) Whole-mount in situ hybridization in the mouse embryo: gene expression in three dimensions. Trends in Genetics **9**, 162–167.
2. Speel E. J. M., Ramaekers F. C. S., and Hopman A. H. N. (1997) Sensitive multicolor fluorescence in situ hybridization using catalyzed reporter deposition (CARD) amplification. J Histochem Cytochem **45**, 1439–1446.
3. Clay H. and Ramakrishnan L. (2005) Multiplex fluorescent in situ hybridization in zebrafish embryos using tyramide signal amplification. Zebrafish **2**, 105–111.
4. Hargrave M., Bowles J., and Koopman P. (2006) In situ hybridization of whole-mount embryos. Methods Mol Biol **326**, 103–113.
5. Lowe L. A. and Kuehn M. R. (2000) Whole mount in situ hybridization to study gene expression during mouse development. Methods Mol Biol **137**, 125–137.
6. Heiskanen M., Peltonen L., and Palotie A. (1996) Visual mapping by high resolution FISH. Trends in Genetics **12**, 379–382.
7. Herrington C. S., Graham A. K., and McGee J. O'D. (1991) Interphase cytogenics J Clin Pathol **44**, 33–38.
8. Correia K. M. and Conlon R. A. (2001) Whole-mount in situ hybridization to mouse embryos. Methods **23**, 335–338.
9. Jowett T., Mancera M., Amores A., and Yan Y. (1996) In situ hybridization to embryo whole mounts and tissue sections: mRNA detection and application to developmental studies. In Situ Hybridization (Clark M., ed.), Chapman and Hall, London, UK, 91–121.
10. Piette D., Hendrickx M., Willems E., Kemp C. R., and Leyns L. (2008) An optimized procedure for whole-mount in situ hybridization on mouse embryos and embryoid bodies. Nature Protocols **3**, 1194–1201.

Chapter 3

Determination of Gene Expression Patterns by In Situ Hybridization in Sections

Kateryna Kyrylkova, Sergiy Kyryachenko, Chrissa Kioussi, and Mark Leid

Abstract

In recent years, in situ RNA hybridization technique has found a widespread application in developmental biology. This method has frequently been used to determine gene expression patterns, which is a first step toward understanding of a gene function. Here, we provide a reliable and sensitive method for in situ RNA hybridization on frozen sections of mouse embryo using digoxigenin-labeled RNA probes. This technique can be used to study gene expression patterns at all stages of odontogenesis.

Key words: In situ RNA hybridization, Frozen sections, Digoxigenin-labeled probe, Alkaline phosphatase–antibody conjugate

1. Introduction

In situ RNA hybridization, also referred as hybridization histochemistry, is a powerful technique that enables the detection of specific RNA sequences in individual cells, tissue sections, intact tissue fragments, and whole-mount embryos (1, 2). In essence, in situ hybridization combines histochemistry with molecular biology, allowing studying gene expression. The principle behind in situ RNA hybridization is that labeled complementary RNA strands (probes) hybridize to target RNA under appropriate conditions. The stable hybrids are formed and can be visualized with a detection system (2).

In situ hybridization was introduced in 1969 and relied heavily upon the use of radioisotopes and autoradiography (3). In recent years, a group of techniques employing non-isotopic labeling

Chrissa Kioussi (ed.), *Odontogenesis: Methods and Protocols*, Methods in Molecular Biology, vol. 887,
DOI 10.1007/978-1-61779-860-3_3, © Springer Science+Business Media, LLC 2012

technology have been developed (2, 4, 5). In the protocol described here, probes are labeled with hapten-substituted ribonucleotide, digoxigenin-11-uridine-5′-triphosphate (DIG-11-UTP), and detected with Fab fragments from an anti-digoxigenin antibody, conjugated with alkaline phosphatase (AP). Chromomeric substrate is used to visualize the signal (6–8).

Because of its specificity and increased number of applications, in situ hybridization is proving particularly important in the field of developmental biology, since a fundamental aspect of development is spatial and temporal expression of genes (2, 9, 10). In situ RNA hybridization can be performed on sectioned and whole-mount embryos. Hybridization on sections is widely used for older and bigger embryos, when penetration of the probe into the whole-mount embryo is problematic. It also allows getting a higher resolution analysis of the gene expression. Although whole-mount embryos can be sectioned, the signal intensity might be decreased depending on labeling and detection systems used for whole-mount in situ hybridization (2, 11).

This chapter describes a routine in situ RNA hybridization in sections protocol that allows the spatiotemporal assessment of gene expression during odontogenesis. The protocol is mainly based on the experience with developing mouse teeth and other organs, but should be applicable to other tissues as well.

2. Materials

Prepare all solutions using ultrapure deionized water and analytical grade reagents. Store all reagents according to manufacturer's instructions or, if not applicable, at room temperature (unless indicated otherwise).

2.1. Components for Preparation of DIG-Labeled Probe

1. DEPC-treated dH_2O (see Note 1).
2. 1.5-ml microfuge tubes.
3. Transcription optimized 5× buffer (Promega).
4. DTT, 100 mM.
5. DIG RNA labeling mix (Roche).
6. Protector RNase inhibitor (Roche).
7. Linearized plasmid (see Note 2).
8. RNA polymerase: T3/T7/Sp6.
9. 1% Agarose gel (see Note 3).
10. RQ1 RNase-free DNase (Promega).
11. 4 M $LiCl_2$.

12. Ethanol.
13. 70% Ethanol in dH_2O.

2.2. Components for Preparation of the Frozen Sections

1. 10-cm Petri dish.
2. Dissection microscope.
3. Dissection tools: Watchmaker's forceps and razor blades.
4. Phosphate-buffered saline (PBS): 137 mM NaCl, 2.7 mM KCl, 10 mM Na_2HPO_4, and 1.76 mM KH_2PO_4 in dH_2O, pH 7.4 (with HCl).
5. 24- and 48-well plates.
6. 4% Paraformaldehyde in PBS (see Note 4).
7. 30% Sucrose in PBS.
8. OCT compound.
9. Disposable embedding molds.
10. Ethanol.
11. Dry ice.
12. Cryostat (Leica), knife holder, glass anti-roll guide, disposable microtome knives, and specimen discs (see Note 5).
13. Micro slides superfrost plus.
14. Hot plate.

2.3. Components for In Situ Hybridization

1. Histology slide tray.
2. Glass staining dish.
3. Slide rack.
4. RNase Away.
5. DEPC-treated dH_2O.
6. PBS.
7. 4% Paraformaldehyde in PBS (see Note 4).
8. Proteinase K, 10 mg/ml stock in 100 mM Tris–HCl (pH 7.5), 10 mM EDTA; store in aliquots at −20°C.
9. Acetylation solution: 0.25% acetic anhydride in 100 mM triethanolamine-HCl (pH 8.0); mix with a stir bar (see Note 6).
10. Formamide.
11. SSC, 20× stock: 3 M NaCl, 0.3 M sodium citrate in DEPC-treated dH_2O, pH 4.5.
12. Hybridization buffer: 50% formamide, 5× SSC, 1% SDS, 50 μg/ml yeast tRNA, 50 μg/ml heparin in DEPC-treated dH_2O (see Note 7).
13. Parafilm M.

14. DIG-labeled probe.
15. Plastic wrap.

2.4. Components for Antibody Reaction and Detection

1. Histology slide tray.
2. SSC, 20× stock.
3. NTT: 0.15 M NaCl, 0.1 M Tris–HCl (pH 8.0), and 0.1% Tween 20.
4. Sheep serum: Heat inactivate at 60°C for 30 min; store in aliquots at −20°C.
5. Blocking buffer: 5% heat-inactivated sheep serum and 2% blocking reagent (Roche) in NTT; store in aliquots at −20°C.
6. Anti-digoxigenin-AP, Fab fragments (Roche).
7. 2 M levamisole; store in aliquots at −20°C.
8. 1 M $MgCl_2$.
9. NTTML: 0.15 M NaCl, 0.1 M Tris–HCl (pH 9.5), 0.1% Tween 20, 50 mM $MgCl_2$, and 2 mM levamisole in dH_2O (see Note 8).
10. BM Purple AP substrate (Roche).
11. 0.2% Glutaraldehyde/4% paraformaldehyde in PBS; store in aliquots at −20°C.
12. Ethanol.
13. 95, 70, and 50% ethanol in dH_2O.
14. Xylene.
15. DPX mounting medium.
16. Micro cover glasses.

3. Methods

Carry out all procedures at room temperature unless otherwise specified.

3.1. Preparation of DIG-Labeled Probe

1. Set up a labeling reaction in a total volume of 20 μl in a 1.5-ml microfuge tube: 9 μl of DEPC-treated dH_2O, 4 μl of transcription 5× buffer, 2 μl of DTT, 2 μl of DIG RNA labeling mix, 1 μl of protector RNase inhibitor, 1 μl of 1 μg/μl linearized plasmid, and 1 μl of RNA polymerase (T3/T7/Sp6). Incubate the reaction at 37°C for 2 h.
2. Examine 1 μl of total volume on 1% agarose gel.
3. Degrade DNA with 1 μl of RNase-free DNase at 37°C for 15 min to remove DNA template (see Note 9).

4. Precipitate RNA with 2 μl of 4 M $LiCl_2$ and 100 μl of −20°C 100% ethanol. Pellet RNA at top speed in a microfuge for 15 min.
5. Discard supernatant, add 100 μl of −20°C 70% ethanol, and pellet RNA at top speed in a microfuge for 2 min.
6. Discard supernatant, dry the pellet at −37°C for 30 min, and resuspend RNA in 100 μl of DEPC-treated dH_2O. Aliquot and store the probe at −20°C.

3.2. Preparation of the Frozen Sections

1. Dissect embryos in ice-cold PBS in a Petri dish. Use dissection microscope, if necessary.
2. Remove heads, transfer them into 24- or 48-well plates, and wash in ice-cold PBS (see Note 10).
3. Fix heads in 4% paraformaldehyde in PBS at 4°C for 2 h–overnight.
4. Wash the heads in PBS at 4°C overnight.
5. Incubate heads in 30% sucrose in PBS at 4°C for 1–2 days (see Note 11).
6. Dip embryo heads into OCT for 1 min.
7. Transfer heads to an embedding mold containing OCT. Orient the heads as desired (see Note 12) and freeze in ethanol, which contains dry ice. Store frozen specimen blocks at −80°C for several months.
8. Attach the block to the specimen disc and cut 8- to 20-μm sections at −20°C in the cryostat (see Notes 5 and 13). Thaw mount sections on the room-temperature micro slides.
9. Dry slides on a 40°C hot plate for 30–60 min. Slides can be placed in a slide box and stored at −80°C for several days.

3.3. In Situ RNA Hybridization

For all steps, use DEPC-treated reagents. Carry out all procedures in a histology slide tray, unless otherwise specified. It is recommended to use two slide trays. It is advisable to wipe gloves, slide tray, rack, and glass staining dish with RNase Away.

1. Wash sections 3× with PBS, 5 min each time (to remove OCT) (see Note 14).
2. Fix sections with 4% paraformaldehyde in PBS for 10 min.
3. Wash the sections 3× with PBS, 3 min each time.
4. Treat sections with 1 μg/ml proteinase K in PBS for 15–20 min (see Note 15).
5. Rinse sections 1× with PBS.
6. Refix sections with 4% paraformaldehyde in PBS for 5 min.
7. Wash as in step 3.

8. Transfer slides to a slide rack in a glass dish with acetylation solution. Acetylate the sections for 10 min (see Note 6).
9. Transfer slides to a slide tray. Wash as in step 1.
10. Transfer slides to a second slide tray, pre-warmed at 68°C and humidified with 50% formamide/5× SSC. Add 200 μl per slide of pre-warmed at 68°C hybridization buffer. Cover the slides with Parafilm coverslips, being careful not to trap any air bubbles. Prehybridize the sections for 2–4 h (see Note 9).
11. Carefully remove Parafilm coverslips and excess buffer from slides. Add 120 μl per slide of pre-warmed at 68°C hybridization buffer, containing the 4 μg/ml of a probe. Cover the slides with Parafilm coverslips carefully. The slide tray should be tightly covered with a lid and wrapped in a plastic wrap. Hybridize the sections in a humidified slide tray at 68°C overnight (see Note 16).

3.4. Antibody Reaction and Detection

Carry out all procedures in a histology slide tray unless otherwise specified.

1. Following hybridization, immerse slides in 5× SSC pre-warmed at 70°C, shake gently, and carefully remove Parafilm coverslips.
2. Incubate sections 2× in pre-warmed 0.2× SSC at 70°C, 30 min each time.
3. Wash sections 1× with 0.2× SSC for 5 min.
4. Wash sections 1× with NTT for 5 min.
5. Add 200 μl per slide of blocking buffer. Cover the slides with Parafilm coverslips carefully. Block the sections for 1–2 h.
6. Carefully remove Parafilm coverslips and excess buffer from the slides. Incubate each slide with 1:5,000 dilution anti-digoxigenin AP fab fragments in blocking buffer. Cover the slides with Parafilm coverslips carefully. Incubate the sections at 4°C overnight.
7. Carefully remove Parafilm coverslips and incubate the sections 3× in NTT, 30 min each time.
8. Wash sections 3× with NTTML, 5 min each time (see Note 8).
9. Incubate sections with BM purple AP substrate in dark until the signal of proper intensity is developed (see Note 17; Fig. 1).
10. Once a signal is at the desired intensity, wash the sections 3× with PBS, 5 min each time.
11. Fix the sections with 0.2% glutaraldehyde/4% paraformaldehyde in PBS for 1–2 h.

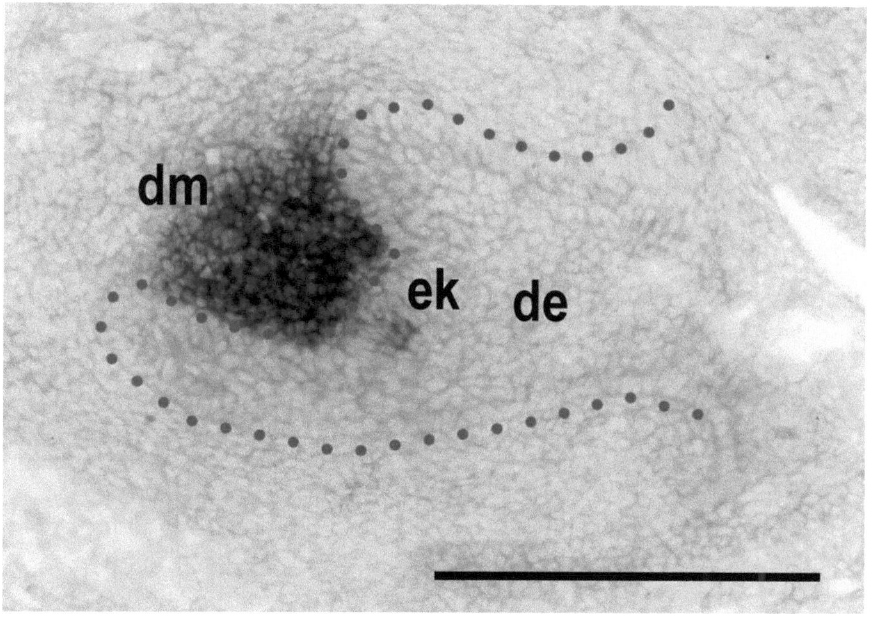

Fig. 1. In situ hybridization analysis of the expression pattern of Fgf-3 in sagittal frozen sections of mouse mandibular incisor at the cap stage (embryonic day 14.5) of tooth development. Signaling molecule Fgf-3 is abundantly expressed in the dental mesenchyme and to a lesser extent in the enamel knot, but not in other parts of the dental epithelium of the developing incisor. Dental enamel is outlined. *ek* enamel knot, *de* dental epithelium, *dm* dental mesenchyme. Scale bar: 200 μm.

12. Rinse sections 3× with PBS.
13. Transfer the slides to a slide rack. Dehydrate sections 1× in 50, 70, and 95% ethanol in dH_2O and 2× in 100% ethanol, 5 min each time.
14. Incubate sections 2× in xylene for 5 min (see Note 18).
15. Mount slides with DPX and apply micro cover glasses, being careful not to trap any air bubbles. Let slides dry overnight (see Note 18).

4. Notes

1. It is important to avoid contamination of the RNA probe by RNase before RNA hybridization. Therefore, it is highly advisable to use DEPC-treated dH_2O to prevent RNase contamination. Add 1 ml of DEPC to 1 l of dH_2O or reagent, shake vigorously, leave overnight at room temperature, and autoclave.
2. Plasmid vectors used as transcription templates should be linearized by restriction enzyme digestion. The DNA should be cut in a way such that RNA polymerase will fall off at the 5′ end

of the gene of interest. The product can be run on 1% agarose gel to ensure that the plasmid is linearized completely. It is recommended to also transcribe sense probe to use it as a negative control.

3. Use DEPC-treated or freshly autoclaved dH_2O to make 1% agarose gel.
4. Make fresh 4% paraformaldehyde each time. Preparation should be carried out inside a fume hood. Store it at 4°C for up to 1 week.
5. See manufacturer's manual for more details.
6. Make fresh acetylation solution each time. 400 ml of solution is enough to fill glass staining dish.
7. It is recommended to make fresh hybridization buffer each time.
8. It is advisable to add $MgCl_2$ and fresh aliquot of levamisole directly before use.
9. DNA degradation is optional.
10. Make sure that most of blood is washed away (blood inhibits fixation). Heads of older embryos can be cut by half sagittaly at midline using razor blade. Each half can be embedded separately.
11. Tissue has to sink in 30% sucrose in PBS.
12. Standard orientation planes for sections include sagittal, transverse, and frontal.
13. Do not place sections on the edges of the slides. During incubation of sections at 68°C, Parafilm M tends to shrink and does not cover the edges of the slide.
14. If the slides were frozen, defrost them for 45 min at room temperature. Make sure that slides do not touch each other or walls of the slide tray.
15. Proteinase K solution should be freshly made each time. The time of treatment with proteinase K solution can be further optimized depending on specimen tissue and section thickness.
16. The concentration of the probe and temperature needed for hybridization will require optimization for each probe and tissue type to assure intense-specific signal and low background.
17. Signal development reaction can take from 1 h to 3–4 days. It is advisable to use sense probe as a negative control and watch background development.
18. The procedure should be carried out inside a fume hood.

References

1. Nilaver G. (1986) *In situ* hybridization histochemistry as a supplement to immunohistochemistry. *In Situ Hybridization in Brain*. New York: Plenum Press.
2. Jin L. and Lloyd R. V. (1997) *In situ* hybridization: methods and applications. *Journal of Clinical Laboratory Analysis* 11, 2–9.
3. Buongiorno-Nardelli M. and Amaldi F. (1969) Autoradiographic detection of molecular hybrids between rRNA and DNA in tissue sections. *Nature* 225, 946–947.
4. John H. L., Birnstiel M. L., and Jones K.W. (1969) RNA-DNA hybrids at the cytological level. *Nature* 223, 912–913.
5. Chan V.T.-W. and McGee J.O'D. (1990) Non-radioactive probes: preparation, characterization, and detection. *In Situ Hybridization: Principle and Practice* (J. M. Polak & J. O'D. McGee eds.) Oxford: Oxford University Press.
6. Holtke HJ, Ankenbauer W, Muhlegger K, Rein R, Sagner G, Seibl R, Walter T. (1995) The digoxigenin (DIG) system for non-radioactive labelling and detection of nucleic acids – an overview. *Cell Mol Biol* 41(7), 883–905.
7. Herrington C. S., Burns J., Graham A. K., Evans M., and McGee J. (1989) Interphase cytogenetics using biotin and digoxigenin labelled probes I: relative sensitivity of both reporter molecules for detection of HPV16 in CaSki cells. *J Clin Pathol* 42, 601–606.
8. Schaeren-Wiemers N. and Gerfin-Moser A. (1993) A single protocol to detect transcripts of various types and expression level in neural tissue and cultured cells: in situ hybridization using digoxigenin-labelled cRNA probes. *Histochemistry* 100, 431–440.
9. Braissant O. and Wahli W. (1998) A simplified *in situ* hybridization protocol using non-radioactively labeled probes to detect abundant and rare mRNAs on tissue sections. *Biochemica* 1, 10–16.
10. Moter A, Göbel UB. (2000) Fluorescence *in situ* hybridisation (FISH) for direct visualization of microorganism. *J Microbiol Meth* 41, 85–112.
11. Nieto M. A., Patel K., and Wilkinson D. G. (1996) *In situ* hybridization analysis of chick embryos in whole mount and tissue sections. *Methods Cell Biol.* 51, 219–235.

Chapter 4

Immunohistochemistry and Detection of Proliferating Cells by BrdU

Sergiy Kyryachenko, Kateryna Kyrylkova, Mark Leid, and Chrissa Kioussi

Abstract

Immunohistochemistry is a classic technique used for the localization of antigenic target molecules in tissue. The method exploits the principle that the target antigen is recognized by specific antibody and is visualized using different detection systems. The subject of this chapter is simultaneous immunohistochemical detection of protein antigens and proliferation marker BrdU in the developing tooth.

Key words: Immunohistochemistry, Mouse embryos, Frozen sections, Protein antigens, BrdU, Proliferation

1. Introduction

Immunohistochemistry (IHC) is currently one of the most popular histological techniques, which is widely used to recognize and detect antigens in different tissues. The principle behind the IHC is specific binding of antibodies to target antigenic epitopes and their subsequent detection (1, 2).

The method was introduced in 1941, when identification of tissue antigens using a direct fluorescence protocol was described (3). Since then, many monoclonal and polyclonal primary antibodies have been developed. The primary antibody can be directly conjugated to a fluorophore or an enzyme, which can catalyze color reaction. Alternatively, the primary antibody can be recognized with labeled species-specific secondary antibodies and detected indirectly. In some systems, such as avidin–biotin complex (ABC), tertiary antibodies are used. Secondary and tertiary antibodies allow amplification of the signal, and subsequently more

Chrissa Kioussi (ed.), *Odontogenesis: Methods and Protocols*, Methods in Molecular Biology, vol. 887,
DOI 10.1007/978-1-61779-860-3_4, © Springer Science+Business Media, LLC 2012

sensitive detection of antigens (2, 4). In this chapter, the primary antibodies are detected by secondary antibodies conjugated to fluorescent cyanine dyes (e.g., Cy-2, Cy-3, and Cy-5) (5).

In 1981, immunostaining based on the use of antibodies against 5-bromo-2′-deoxyuridine (BrdU) was first used to detect cells in S (synthetic) phase of cell cycle. BrdU is a thymidine analog, which can be incorporated into the newly synthesized DNA of replicating cell and recognized by anti-BrdU-specific antibodies. Binding of the antibodies requires DNA denaturation, achieved by heat treatment or low pH. BrdU immunostaining and subsequent quantification of cells positive for BrdU allow determination of proliferation rates and cell cycle kinetics in different tissues (6–8).

In many cases, two or more antigens can be detected in the same tissue specimen at the same time. Multicolor immunohistochemistry that uses double and triple labeling is possible and can be used to study co-localization of several molecules. When two or more antigens are present in the same compartment, their co-localization is marked by a mixed color (9, 10).

In this chapter, we present a technique that has proven effective in the detection of protein antigens and BrdU in the frozen sections of embryonic mouse tooth.

2. Materials

Prepare all solutions using ultrapure deionized water and analytical grade reagents. Store all reagents according to manufacturer's instructions or, if not applicable, at room temperature (unless indicated otherwise).

2.1. In Vivo Labeling of Murine Cells with BrdU

1. 5BrdU.
2. Phosphate-buffered saline (PBS): 137 mM NaCl, 2.7 mM KCl, 10 mM Na_2HPO_4, and 1.76 mM KH_2PO_4 in dH_2O, pH 7.4 (with HCl).
3. BrdU solution: Dissolve 5 mg of BrdU in 1 ml of PBS. Store at −20°C.
4. 1.5-ml microfuge tubes.
5. 1-ml syringes.

2.2. Preparation of the Frozen Sections

1. 10-cm Petri dish.
2. Dissection microscope.
3. Dissection tools: Watchmaker's forceps and razor blades.
4. PBS.
5. 24- and 48-well plates.
6. 4% Paraformaldehyde in PBS (see Note 1).

7. 30% Sucrose in PBS.
8. OCT compound.
9. Disposable embedding molds.
10. Ethanol.
11. Dry ice.
12. Cryostat (Leica), knife holder, glass anti-roll guide, disposable microtome knives, and specimen discs (see Note 2).
13. Micro slides superfrost plus.
14. Hot plate.

2.3. Immunostaining of Protein Antigens

1. Histology slide tray.
2. Parafilm M.
3. Filter paper.
4. 0.2-μm filters.
5. PBS.
6. Methanol.
7. Fetal bovine serum: Heat inactivate at 60°C for 30 min; store in aliquots at −20°C.
8. Donkey serum: Heat inactivate at 60°C for 30 min; store in aliquots at −20°C.
9. Triton X-100.
10. Blocking buffer: 0.3% blocking reagent (Roche), 5% heat-inactivated fetal bovine serum, 5% heat-inactivated donkey serum, and 0.1% Triton X-100 in PBS. Filter through filter paper and then through 0.2-μm filter. Store in aliquots at −20°C.
11. Primary antibody against an antigenic protein of interest.

2.4. Post-antibody Washes and BrdU Immunostaining

1. Histology slide tray.
2. Incubator.
3. Parafilm M.
4. PBS.
5. PBST: PBS containing 0.1% Tween 20.
6. Blocking buffer.
7. Species-specific cyanine-labeled secondary antibodies, which bind primary antibodies against antigenic protein.
8. 4% Paraformaldehyde in PBS (see Note 1).
9. 2 N HCl in dH_2O.
10. Boric acid.
11. 0.1 M borate buffer. Adjust the pH to 8.5 with 1 M NaOH.
12. Anti-BrdU antibodies.

2.5. Post-BrdU-Immunostaining Washes

1. Histology slide tray.
2. Slide rack.
3. PBST.
4. Blocking buffer.
5. Parafilm M.
6. Species-specific cyanine-labeled secondary antibodies, which bind anti-BrdU antibodies.
7. 100, 95, 70, and 50% ethanol in dH_2O.
8. Xylene.
9. DPX mounting medium.
10. Micro cover glasses.

3. Methods

Carry out all procedures at room temperature unless otherwise specified.

3.1. In Vivo Labeling of Mouse Cells with BrdU

1. Inject 1 μl of BrdU solution per 1-g body weight into timed-pregnant mouse (see Note 3).
2. Wait for 2 h before euthanize the animal (see Note 4).

3.2. Preparation of the Frozen Sections

1. Dissect embryos in ice-cold PBS in a Petri dish under dissection microscope, if necessary.
2. Remove heads and transfer them into 24- or 48-well plates, wash in ice-cold PBS (see Note 5).
3. Fix heads in 4% paraformaldehyde in PBS at 4°C for 1–2 h (see Note 6).
4. Wash heads in PBS at 4°C overnight.
5. Incubate heads in 30% sucrose in PBS at 4°C for 1–2 days (see Note 7).
6. Dip heads into OCT for 1 min.
7. Transfer heads to an embedding mold containing OCT. Orient heads as desired (see Note 8) and freeze in ethanol, which contains dry ice. Store frozen specimen blocks at −80°C for several months.
8. Attach block to the specimen disc and cut 8- to 20-μm sections at −20°C in the cryostat (see Note 2). Thaw mount sections at room-temperature micro slides.
9. Dry slides on a 40°C hot plate for 30–60 min. Slides can be placed in a slide box and stored at −80°C for several days.

3.3. Immunostaining of Protein Antigens

Carry out all procedures in a histology slide tray unless otherwise specified.

1. Wash sections 3× with PBS, 10 min each time (to remove OCT).
2. Rinse sections with cold methanol at −20°C for 2–3 min (see Note 9).
3. Add 200 μl per slide of blocking buffer. Cover the slides with Parafilm coverslips carefully and incubate for 1 h.
4. Carefully remove Parafilm coverslips and excess buffer from slides. Add 120 μl per slide of primary antibodies against antigenic protein diluted in blocking buffer. Cover the slides with Parafilm coverslips carefully. Incubate the slides at 4°C overnight (see Note 10).

3.4. Post-antibody Washes and Immunostaining of BrdU

Carry out all procedures in a histology slide tray unless otherwise specified.

1. Wash sections 3× with PBST, 10 min each time.
2. Add 120 μl per slide of blocking buffer with species-specific cyanine-labeled secondary antibodies, which bind primary antibodies against antigenic protein. Cover the slides with Parafilm coverslips carefully. Incubate the slides for 2 h (see Note 10).
3. Wash sections 3× with PBST, 10 min each time.
4. Fix sections with 4% paraformaldehyde in PBS for 5 min.
5. Wash sections 3× with PBS, 5 min each time.
6. Transfer the slide rack to air incubator at 37°C. Add 2 N HCl and incubate the slides for 30 min.
7. Wash sections with 0.1 M borate buffer, pH 8.5, for 15 min.
8. Wash sections 3× with PBS, 5 min each time.
9. Add 200 μl per slide of blocking buffer. Cover the slides with Parafilm coverslips carefully and incubate for 1 h.
10. Carefully remove Parafilm coverslips and excess buffer from the slides. Add 120 μl per slide of anti-BrdU antibodies diluted in blocking buffer. Cover the slides with Parafilm coverslips carefully. Incubate the slides at 4°C overnight (see Note 10).

3.5. Post-BrdU-Immunostaining Washes

Carry out all procedures in a histology slide tray unless otherwise specified.

1. Wash sections 3× with PBST, 10 min each time.
2. Add 120 μl per slide of blocking buffer with species-specific cyanine-labeled secondary antibodies, which bind anti-BrdU. Cover the slides with Parafilm coverslips carefully. Incubate the slides for 2 h (see Note 10).

3. Wash sections 3× with PBST, 10 min each time.
4. Transfer the slides to a slide rack. Wash the sections 2× with dH_2O, 3 min each time.
5. Dehydrate sections 1× in 50, 70, and 95% ethanol in water, and 2× in 100% ethanol, 3 min each time.
6. Incubate sections 2× in xylene for 3 min (see Note 11).
7. Mount slides with DPX and apply micro cover glasses, being careful not to trap any air bubbles. Let slides dry overnight (see Note 11).

4. Notes

1. Make fresh 4% paraformaldehyde each time. Preparation should be carried out inside a fume hood. Store it at 4°C for up to 1 week.
2. See manufacturer's manual for more details.
3. Intraperitoneal injections are recommended for mice.
4. Incorporation of BrdU can be detected in thymus and bone marrow in as little as 1 h post injection and 2 h in tooth. At 24 h post injection, BrdU can be detected in most tissues.
5. Make sure that most of blood is washed away (blood inhibits fixation). Heads of older embryos can be cut by in half sagittaly at midline using razor blade. Each half can be embedded separately.
6. The fixation timing should be optimized for each tissue individually, for example 2 h for E18.5 heads, 1 h 30 min for E16.5 heads, 1 h 15 min for E14.5 heads, etc.
7. Tissue has to sink in 30% sucrose in PBS.
8. Standard orientation planes for sections include sagittal, transverse, and frontal.
9. Permeabilization should only be required for intracellular epitopes when the antibody requires access to the inside of the cell to detect the protein. However, it will also be required for detection of transmembrane membrane proteins if the epitope is in the cytoplasmic region. There are a wide variety of detergents to choose from, and they differ in their efficiency to extract membranes. The type of detergent and time for the permeabilization will require optimization for the tissue of interest.
10. Correct dilutions will contribute to the quality of staining if they are prepared accurately and consistently. Often, a manufacturer recommends dilution ranges compatible with other

variables, such as method, incubation time, and temperature. If this information is not provided, optimal working dilution of the antibodies must be determined by titration.

11. The procedure should be carried out inside a fume hood.

References

1. Ramos-Vara J. A. (2005) Technical aspects of immunohistochemistry. *Vet Pathol* 42(4), 405–26.
2. Buchwalow I. B. and Bocker W. (2010) *Immunohistochemistry: Basics and Methods,* 1st ed. Springer, p. 153.
3. Coons A. H., Creech H. J., Jones R. N. (1941) Immunological properties of an antibody containing a fluorescent group. *Proc Soc Exp Biol Med* 47, 200–202.
4. Bergroth V. (1983) Comparison of various immunohistochemical methods. *Histochemistry* 77, 177–184.
5. Waggoner A. (1997) Cyanine dyes as labelling reagents for detection of biological and other materials by luminescence methods. *US Patent* No. 5627027.
6. Shimada A, Shibata T, Komatsu K, and Nifuji A. (2008) Improved methods for immunohistochemical detection of BrdU in hard tissue. *J Immunol Methods* 339(1), 11–16.
7. Dolbeare F. (1995) Bromodeoxyuridine: a diagnostic tool in biology and medicine, Part I: historical perspectives, histochemical methods and cell kinetics. *Histochem J* 27, 339–369.
8. Gratzner H.G. (1982) Monoclonal antibody to 5-bromo- and 5-iododeoxyuridine: a new reagent for detection of DNA replication. *Science* 218, 474–475.
9. Nakane P. K. (1968) Simultaneous localization of multiple tissue antigens using the peroxidase-labeled antibody method: a study on pituitary glands in the rat. *J Histochem Cytochem* 16, 557–559.
10. Tsurui H., Nishimura H., Hattori S., Hirose S., Okumura K., and Shirai T. (2000) Seven-color fluorescence imaging of tissue samples based on Fourier spectroscopy and singular value decomposition. *J Histochem Cytochem* 48(5), 653–662.

Chapter 5

Detection of Apoptosis by TUNEL Assay

Kateryna Kyrylkova, Sergiy Kyryachenko, Mark Leid, and Chrissa Kioussi

Abstract

Terminal deoxynucleotidyl transferase (TdT) dUTP Nick-End Labeling (TUNEL) assay has been designed to detect apoptotic cells that undergo extensive DNA degradation during the late stages of apoptosis. The method is based on the ability of TdT to label blunt ends of double-stranded DNA breaks independent of a template. This chapter describes an assay for detection of apoptotic cells during mouse odontogenesis using a colorimetric TUNEL system.

Key words: Apoptosis, TUNEL, Immunofluorescence, Mouse embryos, Frozen sections

1. Introduction

Apoptosis is the process of programmed cell death that occurs under normal physiological conditions, such as embryogenesis, tissue homeostasis, and immune system regulation, and can be induced by various physical and chemical stimuli. Cells undergoing apoptosis show characteristic morphological and biochemical features, which include chromatin condensation, cell and nuclear shrinkage, formation of membrane-bound cell fragments, known as apoptotic bodies, and rapid phagocytosis by neighboring cells or macrophages without associated inflammation. The biochemical hallmark of apoptosis is degradation of DNA by endonucleases, which produce double-stranded oligonucleosomal DNA fragments (1–4). These DNA fragments are 180–200 bp in size and can be separated into a ladder-like pattern on agarose gel electrophoresis (4). However, this method cannot provide information regarding the histological localization of DNA fragmentation at a single-cell level or in mixed cell populations.

Chrissa Kioussi (ed.), *Odontogenesis: Methods and Protocols*, Methods in Molecular Biology, vol. 887,
DOI 10.1007/978-1-61779-860-3_5, © Springer Science+Business Media, LLC 2012

Terminal deoxynucleotidyl transferase (**T**dT) Terminal deoxynucleotidyl transferase (**T**dT) d**U**TP **N**ick-**E**nd **L**abeling (TUNEL) is an assay for localization of apoptotic DNA fragmentation in situ that was originally described in 1992 (5). The method relies on the template-independent identification of blunt ends of double-stranded DNA breaks by TdT. The enzyme catalyzes the addition of labeled dUTPs to a 3′-hydroxyl termini of DNA ends, which can be visualized using immunohistochemical techniques (6, 7).

The staining kinetics of the TUNEL assay depends on reagent concentration, fixation of the tissue, extent of proteolysis, and accessibility of DNA strand breaks, which vary between tissue types (1, 2). Therefore, it is important to standardize the technique by using tissue sections with DNAse treatment as a positive control and without TdT treatment as a negative control of apoptosis in order to avoid false-positive or -negative results.

One has to keep in mind that DNA damage is not a unique feature of apoptosis, but can also occur in necrosis. Therefore, the accuracy of the TUNEL assay as a method to detect apoptosis has been questioned in several studies (8, 9). Thus, it might be important to use another independent method, along with the TUNEL assay, to confirm and characterize apoptosis. Such methods include immunohistochemical staining for apoptosis-induced protease caspase 3, Western blots of PARP cleaved by caspases, detection of phosphatidylserine on the cell surface with Annexin V, etc. (7).

In this chapter, we describe a method for detection of apoptosis in the frozen sections of embryonic dental tissues using DeadEnd^{tm} colorimetric TUNEL system (Promega). Biotinylated dUTPs were recognized by streptavidin–Cyanine 3 (SA–Cy3) conjugate, which yielded the best specificity for us. Also, this technique allows TUNEL and immunofluorescence double labeling for apoptotic cells with specific antigen(s), which can be detected by other cyanine-conjugated antibodies, if needed (10, 11).

2. Materials

Prepare all solutions using ultrapure deionized water and analytical grade reagents. Store all reagents according to manufacturer's instructions or, if not applicable, at room temperature (unless indicated otherwise).

2.1. Components for Preparation of the Frozen Sections

1. 10-cm Petri dish.
2. Dissection microscope.
3. Dissection tools: Watchmaker's forceps and razor blades.

4. Phosphate-buffered saline (PBS): 137 mM NaCl, 2.7 mM KCl, 10 mM Na_2HPO_4, and 1.76 mM KH_2PO_4 in dH_2O, pH 7.4 (with HCl).
5. 24- and 48-well plates.
6. 4% Paraformaldehyde in PBS (see Note 1).
7. 30% Sucrose in PBS.
8. OCT compound.
9. Disposable embedding molds.
10. Ethanol.
11. Dry ice.
12. Cryostat (Leica), knife holder, glass anti-roll guide, disposable microtome knives, and specimen discs (see Note 2).
13. Micro slides superfrost plus.
14. Hot plate.

2.2. Components for TUNEL Assay

Equilibration buffer, biotinylated nucleotide mix, and recombinant terminal deoxynucleotidyl transferase (rTdT) are components of DeadEnd[tm] colorimetric TUNEL system (Promega).

1. Histology slide tray.
2. Slide rack.
3. Incubator.
4. Plastic coverslips.
5. PBS.
6. 4% Paraformaldehyde in PBS.
7. Proteinase K, 10 mg/ml stock in 100 mM Tris–HCl (pH 7.5), and 10 mM EDTA; store in aliquots at 20°C.
8. Equilibration buffer.
9. TdT reaction mix: Add 98 μm of equilibration buffer, 1 μm of biotinylated nucleotide mix, and 1 μm of rTdT per one reaction (see Note 3).
10. Negative control mix: Add 98 μm of equilibration buffer, 1 μm of biotinylated nucleotide mix, and 1 μm of autoclaved dH_2O per one reaction.
11. SSC, 20× stock: 3 M NaCl and 0.3 M sodium citrate in dH_2O, pH 4.5.
12. PBST: PBS containing 0.1% Tween 20.
13. SA–Cy3 (Jackson ImmunoResearch) (see Note 4).
14. 100, 95, 70, and 50% ethanol in dH_2O.
15. Xylene.
16. DPX mounting medium.
17. Micro cover glasses.

2.3. Components for Preparation of TUNEL Positive Control

1. Histology slide tray.
2. Plastic coverslips.
3. PBS.
4. DNase I buffer: 40 mM Tris–HCl (pH 7.9), 10 mM NaCl, 6 mM $MgCl_2$, 10 mM $CaCl_2$ in dH_2O.
5. DNase I.

3. Methods

Carry out all procedures at room temperature unless otherwise specified.

3.1. Preparation of the Frozen Sections

1. Dissect the embryos in ice-cold PBS in a Petri dish under dissection microscope, if necessary.
2. Remove heads and place them into a 24- or 48-well plates, wash with ice-cold PBS (see Note 5).
3. Fix heads in 4% paraformaldehyde in PBS at 4°C for 2 h–overnight.
4. Wash heads in PBS at 4°C overnight.
5. Incubate heads in 30% sucrose in PBS at 4°C for 1–2 days (see Note 6).
6. Dip heads into OCT for 1 min.
7. Transfer embryo heads to an embedding mold containing OCT. Orient heads as desired (see Note 7) and freeze in ethanol, which contains dry ice. Store frozen specimen blocks at −80°C for several months.
8. Attach the block to the specimen disk and cut 8- to 20-μm sections at −20°C in the cryostat (see Note 2). Thaw mount sections on the room-temperature micro slides. Prepare additional slides for positive and negative controls, if needed.
9. Dry slides on a 40°C hot plate for 30–60 min. Slides can be placed in a slide box and stored at −80°C for several days.

3.2. TUNEL Assay

Carry out all procedures in a histology slide tray unless otherwise specified.

1. Wash sections 2× with PBS, 5 min each time (to remove OCT).
2. Fix sections with 4% paraformaldehyde in PBS for 15 min.
3. Wash the sections 2× with PBS, 5 min each time.
4. Treat sections with 1 μg/ml proteinase K in PBS for 10–20 min (see Note 8).

5. Wash sections with PBS for 5 min.
6. Refix sections with 4% paraformaldehyde in PBS for 5 min.
7. Wash sections 2× with PBS, 5 min each time.
8. Prepare a positive control (optional, see Subheadings 2.3 and 3.3).
9. Remove excess liquid from the slides. Add 100 μl per slide of equilibration buffer. Cover slides with plastic coverslips carefully and incubate for 5–10 min.
10. Carefully remove plastic coverslips and excess buffer from the slides. Add 100 μl per slide of TdT reaction mix. Meanwhile, prepare a negative control (optional): Add negative control mix instead of TdT reaction mix. Cover the slides with plastic coverslips carefully. Incubate the slides at 37°C for 60 min.
11. Wash sections with 2× SSC for 15 min.
12. Wash sections 3× with PBS, 5 min each time.
13. Add 120 μl per slide of SA–Cy3 diluted in PBS (1:500). Cover the slides with plastic coverslips carefully. Incubate the slides for 30–45 min (see Note 9).
14. Wash sections 3× with PBST, 10 min each time.
15. Transfer slides to a slide rack. Wash the sections 2× with dH_2O, 3 min each time.
16. Dehydrate sections 1× in 50, 70, and 95% ethanol in water, and 2× in 100% ethanol, 3 min each time.
17. Incubate sections 2× in xylene, 3 min each time (see Note 10).
18. Mount slides with DPX and apply micro cover glasses, being careful not to trap any air bubbles. Let slides dry overnight (see Note 10).

3.3. Preparation of TUNEL Positive Control

Carry out all procedures in a histology slide tray unless otherwise specified.

1. Add 120 μl per slide of DNase I buffer. Cover slides with plastic coverslips carefully. Incubate the slides for 5 min.
2. Carefully remove plastic coverslips and excess of buffer from the slides. Add 100 μl per slide of DNase I buffer containing 5–10 units/ml of DNase I. Cover slides with plastic coverslips carefully. Incubate the slides for 10 min (see Note 11).
3. Wash sections 3× with dH_2O, 5 min each time.
4. Wash sections with PBS for 5 min. Process the positive control as described in Subheading 3.2, step 8.

4. Notes

1. Make fresh 4% paraformaldehyde each time. Preparation should be carried out inside a fume hood. Store it at 4°C for up to 1 week.
2. See manufacturer's manual for more details.
3. Prepare sufficient reaction mix for all experimental and control reaction.
4. Any other kind of labeled streptavidin can be used to detect biotinylated dUTP.
5. Make sure that most of blood is washed away (blood inhibits fixation). Heads of older embryos can be cut by half sagittaly at midline using razor blade. Each half can be embedded separately.
6. Tissue has to sink in 30% sucrose in PBS.
7. Standard orientation planes for sections include sagittal, transverse, and frontal.
8. Proteinase K solution should be freshly made each time. Timing is important for treatment with proteinase K and should be adjusted for every tissue type. Proteinase K concentration of 1 μg/ml worked best for treatment of dental tissues. Higher concentrations of proteinase K cause dental epithelium to come off the slide. Alternatively, the sections can be treated with 100% methanol for 2–3 min to permeabilize the tissue.
9. Correct dilutions will contribute to the quality of staining if they are prepared accurately and consistently. Often, a manufacturer recommends dilution ranges compatible with other variables, such as method, incubation time, and temperature. If this information is not provided, optimal working dilution of the antibodies must be determined by titration.
10. The procedure should be carried out inside a fume hood.
11. An optimization step may be required when using other kinds of DNases.

References

1. Saraste A. (1999) Morphologic criteria and detection of apoptosis. *Herz* 24 (3), 189–195.
2. Saraste A., Pulkki K. (2000) Morphologic and biochemical hallmarks of apoptosis. *Cardiovasc Res* 45 (3), 528–537.
3. Cohen J. J. (1993) Apoptosis. *Immunol Today* 14 (3), 126–130.
4. Bortner C. D., Oldenburg N. B., Cidlowski J. A. (1995) The role of DNA fragmentation in apoptosis. *Trends Cell Biol* 5 (1), 21–26.
5. Gavrieli Y., Sherman Y., Ben-Sasson S. A. (1992) Identification of programmed cell death in situ via specific labeling of nuclear DNA fragmentation. *J Cell Biol* 119 (3), 493–501.

6. Nagoescu A., Lorimier P., Labat-Moleur F., Drouet C., Robert C., Guillermet C., Brambilla C., Brambilla E. (1996) In situ apoptotic cell labeling by the TUNEL method: improvement and evaluation on cell preparations. *J Histochem Cytochem* 44 (9), 959–968.
7. Rode H. D., Eisel D., Frost I. (2004) Apoptosis, cell death and cell proliferation. 3rd ed. London: Roche Applied Science.
8. Grasl-Kraupp B., Ruttkay-Nedecky B., Koudelka H. Bukowska K., Bursch W., Schulte-Hermann R. (1995) In situ detection of fragmented DNA (TUNEL assay) fails to discriminate among apoptosis, necrosis, and autolytic cell death: a cautionary note. *Hepatology* 21 (5), 1465–1468.
9. Orita Y., Nishizaki K., Sasaki J., Kanda S., Kimura N., Nomiya S., Yuen K., Masuda Y. (1999) Does TUNEL staining during peri- and post-natal development of the mouse inner ear indicate apoptosis? *Acta Otolaryngol Suppl* 540, 22–26.
10. Tornusciolo D. R., Schmidt R. E., Roth K. A. (1995) Simultaneous detection of TDT-mediated dUTP-biotin nick end-labeling (TUNEL)-positive cells and multiple immunohistochemical markers in single tissue sections. *Biotechniques* 19 (5), 800–805.
11. Oberhaus S. M. (2003) TUNEL and immunofluorescence double-labeling assay for apoptotic cells with specific antigen(s). *Methods Mol Biol* 218, 85–96.

Chapter 6

Use of siRNA in Dental Tissue-Derived Cell Cultures: Integrin Knockdown in Fibroblasts

Malgorzata M. Barczyk, Donald Gullberg, and Anne Isine Bolstad

Abstract

Short (or small) interfering RNAs (siRNAs) are double-stranded RNA molecules about 21–25 nucleotides long that have the capacity to disrupt the activity of genes on a posttranscriptional level. This sequence homology-driven gene silencing capacity has been utilized by researchers to selectively block the translation of mRNA to proteins in order to study specific gene functions and identify target molecules. Importantly, siRNAs have the potential to be used in treatment of disease. Here, we describe how the siRNA technology can be used to knock down genes in dental tissue-derived cells using integrin α11 knockdown as an example.

Key words: siRNA, Dental cultures, Integrins, Periodontal ligament, Fibroblasts

1. Introduction

In 1998, Fire et al. (1) showed that introduction of double-stranded RNAs (dsRNAs) into *Caenorhabditis elegans* resulted in disrupted activity of genes with homologous sequences, a discovery for which they were awarded the Nobel Prize in Physiology or Medicine in 2006 (2). The RNA interference (RNAi) occurs post-transcriptionally in animals, but can arise transcriptionally in plants (3, 4). In principle, when dsRNAs are injected into cells, they are cleaved by an endonuclease called Dicer into short interfering RNAs (siRNAs), 21–23 bp long, with two overhanging nucleotides at the 3′ terminal end of each strand (4, 5). The siRNA is unwound by a helicase before entering the RNA-induced silencing complex (RISC), where it binds to the complementary mRNA strand by base pairing, and afterwards the target mRNA strand is

Chrissa Kioussi (ed.), *Odontogenesis: Methods and Protocols*, Methods in Molecular Biology, vol. 887, DOI 10.1007/978-1-61779-860-3_6, © Springer Science+Business Media, LLC 2012 degraded (6). Thus, the translation of target mRNA into proteins is prevented (7). In this chapter, we describe the mechanism of siRNA silencing with synthesized target-specific short oligonucleotides ready to bind RISC complex directly (see Fig. 1).

The biological phenomenon of RNAi has formed the basis for methods developed for artificial posttranscriptional gene silencing. Knocking down gene expression and silencing genes associated with disease by use of siRNA may also have a potential in treatment of different diseases. Thus, the applications of RNAi are several, and in vivo and in vitro applications of siRNA have recently been reviewed (8, 9). Web resources for proper siRNA design are available (9). Predesigned and validated siRNAs and siRNA libraries can also be purchased from different companies.

We here report the present strategies for knocking down target genes in cells from dental tissues by siRNA technology using integrin α11 as an example of a gene target (see Fig. 2). Human gingival

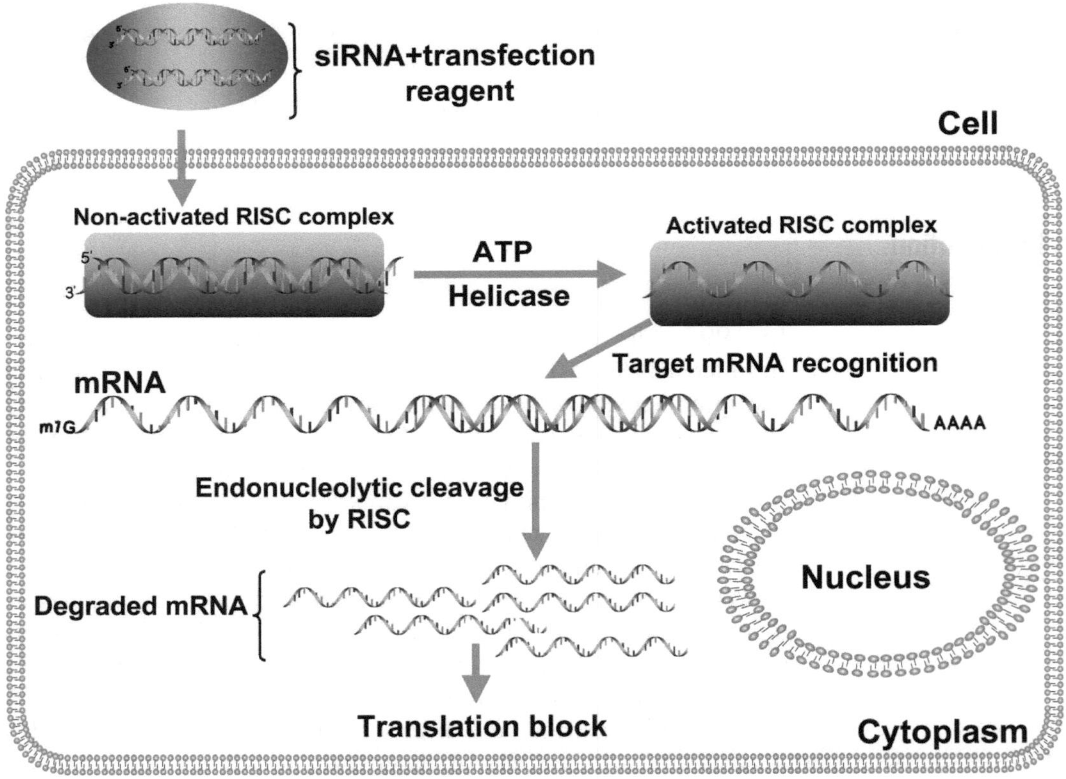

Fig. 1. The mechanism of targeting mRNA with synthetic, sequence-specific siRNA. The siRNAs (19–25 bp long) are bound by transfection reagent, enter the cell, and bind to RNA-induced silencing complex (RISC) (nonactivated). The siRNA strands are unwound by helicase and the strand complementary to the target mRNA remains within RISC (activated). Activated RISC is translocated to mRNA, and the strand complementary to mRNA binds to mRNA. An endonuclease (Argonaute) present in the RISC complex degrades targeted mRNAs.

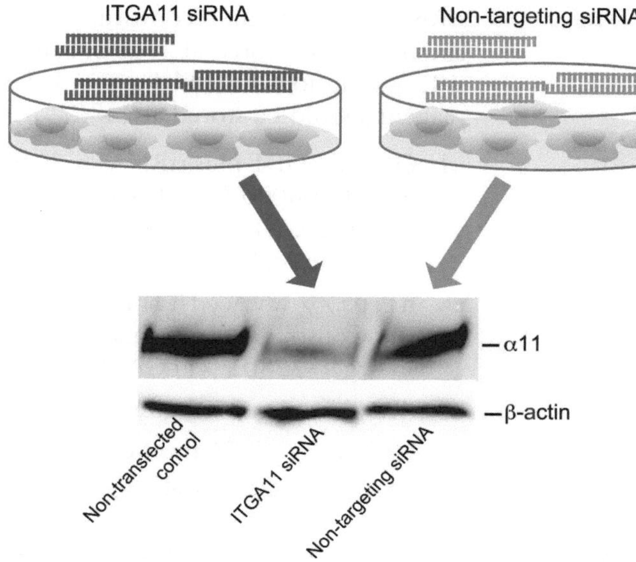

Fig. 2. Silencing of integrin α11 using target-specific siRNA. Cells were transfected with 19-nucleotide-long siRNA targeting human *ITGA11* mRNA and non-targeting (scrambled) siRNA. Cells treated with *ITGA11*-specific siRNA have shown 80–90% reduction in protein level of integrin α11 chain (*middle lane*).

and periodontal ligament (PDL) fibroblasts were transfected with siRNA to human integrin α11. Efficiency of a knockdown was estimated at protein level by western blotting. Functional effect of gene silencing was evaluated in collagen gel contraction assay (10). Thus, the siRNA method can be used successfully in combination with dental tissue-derived cells.

2. Materials

2.1. Cells and Cell Culture

1. Gingival fibroblasts (10, 11).
2. PDL fibroblasts (10, 12).
3. Cell culture flasks.
4. Cell culture 12-well plates.
5. Dulbecco's modified Eagle's medium (DMEM).
6. Fetal calf serum (FCS).
7. Penicillin and streptomycin (PeSt).

2.2. siRNA

siRNAs should be reconstituted in siRNA suspension buffer, aliquoted, and stored at −70°C. Multiple freezing and thawing of siRNA solutions should be avoided (see Note 1). Transfection reagent should be stored at +4°C.

1. ON-TARGET plus siRNAs to human integrin α11 (J-008000-07, J-008000-08, J-008000-09, and J-008000-10) and ON-TARGET plus Non-Targeting siRNA (D-001810-02-05) (see Note 2); link:

   http://www.dharmacon.com/CatalogSearch/ConsolidatedSearch.aspx?searchTerm=ITGA11&searchTarget=22801_+ITGA11.
2. siRNA suspension buffer.
3. HiPerFect siRNA transfection reagent.
4. RNase Away solution.

2.3. Sodium Dodecyl Sulfate-PAGE

1. 2.0 M Tris–HCl, pH 8.8: 121.14 g of Tris base in 500 ml of milli-Q H_2O. pH was adjusted using 12 M HCl.
2. 0.5 M Tris–HCl, pH 6.8: 30.28 g of Tris base in 500 ml of milli-Q H_2O. pH was adjusted using 12 M HCl.
3. 10% ammonium persulfate (APS) solution in milli-Q H_2O.
4. 20% sodium dodecyl sulfate (SDS) solution in milli-Q H_2O.
5. *N,N,N,N'* tetramethylethylenediamine (TEMED).
6. 30% Acrylamide/Bis solution, 37.5:1.
7. 10× Tris/glycine/SDS buffer for SDS-PAGE (formulation for 1× buffer: 25 mM Tris, 192 mM glycine, 0.1% SDS), pH 8.3.
8. DL-Dithiothreitol (DTT).

2.4. Immunotransfer and Immunodetection

1. Hybond-ECL nitrocellulose membrane (code RPN2020D, 0.45 μm, 20×20 cm).
2. Tris-buffered saline Tween-20 (TBS-T), pH 7.4: 8.8 g NaCl, 0.2 g KCl, 3.0 g Tris base, and 500 μl of Tween-20 in total of 1,000 ml of distilled H_2O.
3. Dried skimmed milk.
4. ECL Western Blotting Detection Reagents.

2.5. Antibodies

1. Rabbit anti-human α11 integrin antibody (noncommercial (13)). The antibody can be replaced by commercially available human/mouse integrin α11 antibodies, for example monoclonal rat IgG1 clone # 396214 (R&D Systems) (see Note 3).
2. Monoclonal anti-β-actin antibody produced in mouse, clone AC-15, ascites fluid (Sigma, St. Louis, MO, USA).
3. Secondary antibodies: Goat anti-rabbit IgG-horseradish peroxidase (HRP) (sc-2004) and goat anti-mouse IgG-HRP (sc-2055) (Santa Cruz Biotechnology, Santa Cruz, CA, USA).

3. Methods

3.1. Cells

Primary human PDL fibroblasts and gingival fibroblasts were obtained from tissue explants of healthy molar PDL and gingival tissue (10–12). Cells were cultured in DMEM supplemented with 10% FCS and 1% PeSt at +37°C and 5% CO_2. The cell cultures were split 1:2 once a week by trypsinization (see Note 4).

3.2. siRNA Work

All procedures were performed according to Dharmacon and Qiagen protocols for siRNA reconstitution and siRNA transfection, respectively.

Links to protocols:

http://www.dharmacon.com/empty.aspx?id=2119&imageid=1960

http://www.qiagen.com/products/transfection/transfectionreagents/hiperfecttransfectionreagent.aspx

Details of procedures based on manufacturer's protocols performed in our laboratory are provided below.

3.3. Reconstitution of siRNAs

1. Reconstitution of siRNAs should be performed in sterile hood. All surfaces and pipettes must be decontaminated with RNase Away solution (see Note 5).
2. Lyophilized siRNAs were reconstituted in RNase-free siRNA suspension buffer to 20 μM at room temperature by careful pipetting.
3. Reconstituted siRNA solutions were divided into 10–20-μl aliquots and frozen at −70°C (see Note 1).

3.4. Delivery of siRNA into Human PDL Fibroblasts and Gingival Fibroblasts: Small Scale (see Notes 7 and 8)

1. Human PDL fibroblasts or gingival fibroblasts were seeded into 12-well plate at 10^5 cells and cultured overnight in DMEM supplemented with 1% PeSt and 10% FCS at +37°C (see Note 6).
2. Cell culture medium was changed next day and 1 ml was left in each well 1 h before transfection (see Note 7).
3. 100 μl of DMEM (without antibiotics or FCS) was mixed with 6 μl of HiPerFect transfection reagent, 10 μl of 20 μM stock of siRNA (ITGA11: J-008000-07, J-008000-08, J-008000-09, or J-008000-10 and Non-Targeting siRNA: D-001810-02-05), and incubated for 15 min at room temperature. Transfection toxicity control was included (complete cell culture medium with transfection reagent (Mock transfection)) (see Notes 8, 9, and 10). Final concentration of siRNA in each well was 200 nM.
4. 100 μl of transfection mixture was added carefully into each well and incubated with cells for 36–48 h at +37°C and 5% CO_2.
5. Cells were harvested after incubation and efficiency of the siRNA was tested by western blotting (see Note 11 and Subheading 3.6 of this chapter).

3.5. Delivery of siRNA into Human PDL Fibroblasts and Gingival Fibroblasts: Preparative and Analytical Scale

1. Human PDL fibroblasts or gingival fibroblasts were seeded into a 75-cm^2 flask and cultured until 50% confluence in DMEM supplemented with 1% PeSt and 10% FCS.
2. Cell culture medium was changed next day and 10 ml was left in each flask 1 h before transfection.
3. 1 ml of DMEM (without antibiotics or FCS) was mixed with 40 μl of HiPerFect transfection reagent, 100 μl of 20 μM stock of siRNA (J-008000-10), and incubated for 15 min at room temperature. Final concentration of siRNA was 200 nM.
4. 1 ml of transfection mixture was added carefully into each flask and incubated with cells for 36–48 h at +37°C and 5% CO_2.
5. Cells were harvested after incubation and used for protein detection by western blotting and in functional assays.

3.6. Protein Analysis After ITGA11 siRNA Treatment

3.6.1. Sample Preparation for Western Blotting

Harvested cells were washed in PBS and lysed with 200 μl of sample buffer with 50 mM DTT per well of a 12-well plate.

3.6.2. Sodium Dodecyl Sulfate-PAGE

1. 6% SDS-polyacrylamide gel preparation:
   - *Stacking gel (9 ml).* 6 ml of milli-Q H_2O, 1.5 ml of 0.5 M Tris–HCl, pH 6.8, 45 μl of 10% APS, 45 μl of 20% SDS, 10 μl of TEMED, and 1.5 ml of 30% acrylamide/Bis solution, 37.5:1.
   - *Resolving gel (20 ml).* 12 ml of milli-Q H_2O, 4 ml of 2.0 M Tris–HCl, pH 8.8, 100 μl of 10% APS, 100 μl of 20% SDS, 20 μl of TEMED, and 4 ml of 30% acrylamide/Bis solution, 37.5:1.

   Volumes of gels are estimated for two mini-protean gels (BioRad) of 1.5 mm thickness and 10-well combs.
2. 50 μl of each sample was loaded into the 6% SDS-polyacrylamide gel, and run for 90 min at 100 V.

3.6.3. Immunotransfer

1. Proteins were transferred to cellulose membrane by wet transfer or for 1 h at 100 V.
2. Membranes were blocked for 1 h at room temperature with 5% skimmed milk solution in TBS-T (see Note 12), followed by incubation with primary anti-human α11 antibody (serum diluted 1:500) and anti-β-actin antibody (ascites fluid diluted 1:5,000) (see Note 13) diluted in 1% skimmed milk solution in TBS-T for 2 h at room temperature (see Note 14).
3. Membranes were washed 3×10 min with TBS-T and incubated with secondary goat anti-rabbit IgG-HRP and goat

anti-mouse IgG-HRP antibodies (original stocks of 200 μg/0.5 ml diluted 1:5,000) for 1 h at room temperature.

4. Membranes were developed using ECL Western Blotting Detection Reagents and visualized using ChemiDoc XRS device and quantity One 1D Analysis Software (BioRad) (see Fig. 2).

4. Notes

1. It is important to avoid multiple freeze–thaw cycles when handling siRNA. Aliquots should be planned so that all the content is used in one or two experiments. Multiple freezing and thawing cause degradation of siRNA.
2. In this chapter, we describe the procedure details for siRNAs to human and mouse integrin α11 from Dharmacon only. The siRNAs from other producers were not tested.
3. Antibodies to α11 integrin chain (see Fig. 2): A noncommercial rabbit anti-human α11 integrin antibody was used to detect human α11 integrin (13). There are few anti-human α11 integrin antibodies commercially available on the market. We successfully tested in western blotting only one of them: human/mouse integrin α11 antibody monoclonal rat IgG1 clone # 396214 (R&D Systems).
4. This chapter describes siRNA technique for adherent cells only.
5. While handling siRNA, it is important to work in sterile and RNase-free conditions at all times. Remember to decontaminate the hood and pipettes with RNase Away solution. Make sure that pipette tips and tubes for aliquoting are RNase free. Use RNase-free filter tips only.
6. It is important to have appropriate cell density before transfection. It should be adjusted to cell population doubling time and time of harvesting cells for the readout assay. Make sure that cells are not very dense at the final time point. It can result in loosening of the cell film from the plate. In case of slowly growing cells such as human PDL fibroblasts or gingival fibroblasts (<48 h of population doubling time), cells can be seeded at 80–90% confluence for both PCR and western blotting as readout assays.

   Make sure that cells before seeding are in a single cell suspension and do not form clumps since this can also affect the efficiency of transfection.
7. Cell culture media should be replaced with fresh one 1 h before transfection. Fresh medium will stimulate cells for better uptake

of siRNA and provide better environment for the reaction in the presence of cell-toxic transfection reagent.

8. Testing of new siRNA: Prepare transfection mixtures containing from 2 to 200 nM of siRNA and transfect your target cells as described in Subheading 3. This will allow to find optimal conditions for particular siRNA and cell type. In general, cell type and/or siRNA specificity can affect the efficiency of its silencing effects. Testing mouse *Itga11* siRNA on mouse PDL fibroblasts showed that maximum effect can be achieved at 100 nM of siRNA and the efficiency was about 50% and increasing siRNA concentration did not improve siRNA efficiency. Under the same conditions, human *ITGA11* siRNA blocked the expression of α11 integrin chain almost completely.
9. Further improvement of transfection efficiency: In some cases, adjusting of transfection reagent concentration might be required. One can start from 2× lower (if the transfection reagent is toxic for particular cells) and 2× higher than recommended by the producer. Cell starvation in serum-free condition for 2–3 h before transfection might increase the efficiency of transfection.
10. Controls should be included at all times (see Fig. 2). When planning the plate setup, make sure that you have the following controls with your cells:
    - Cells only with cell culture medium.
    - Cells with cell culture medium + transfection reagent only (mock transfection, toxicity control).
    - Cells with cell culture medium + non-targeting (scrambled) siRNA (siRNA specificity control).
    - Optional—Cells with cell culture medium + siRNA to housekeeping gene (expl. GAPDH) as your transfection efficiency control.
11. The time for testing efficiency of the siRNA is dependent on readout assay. PCR testing requires ca. 18–24 h while western blotting should be performed after 36–48 h. In case of mouse *Itga11* siRNA, no effect was seen on protein level after 24 h. If longer cell culture is required, one can retransfect cells with siRNA after ca. 72 h.
12. 5% Fat-free skimmed milk or 5% solution of BSA in TBS-T can be used to block unspecific sites of nitrocellulose membrane.
13. Primary and secondary antibodies can be diluted in 1% fat-free skimmed milk solution in TBS-T or 1% BSA solution in TBS-T.
14. Incubation conditions in Western blotting for most primary antibody protocols indicate 2 h at +37°C.

Acknowledgments

This work was supported by University of Bergen, Bergen, Norway. The Research Council of Norway (183258/S10) and The Norwegian Cancer Society (536711) M.M. Barczyk is a recipient of grant from Helse Vest, Norway (project number: 911584).

References

1. Fire, A., Xu, S., Montgomery, M. K., Kostas, S. A., Driver, S. E., and Mello, C. C. (1998) Potent and specific genetic interference by double-stranded RNA in *Caenorhabditis elegans*. *Nature* 391, 806–11.
2. http://nobelprize.org/nobel_prizes/medicine/laureates/2006/adv.html.
3. Montgomery, M. K., Xu, S., and Fire, A. (1998) RNA as a target of double-stranded RNA-mediated genetic interference in *Caenorhabditis elegans*. *Proc Natl Acad Sci USA* 95, 15502–7.
4. Bernstein, E., Caudy, A. A., Hammond, S. M., and Hannon, G. J. (2001) Role for a bidentate ribonuclease in the initiation step of RNA interference. *Nature* 409, 363–6.
5. Hammond, S. M., Bernstein, E., Beach, D., and Hannon, G. J. (2000) An RNA-directed nuclease mediates post-transcriptional gene silencing in Drosophila cells. *Nature* 404, 293–6.
6. Kawamata, T., and Tomari, Y. (2010) Making RISC. *Trends Biochem Sci* 35, 368–76.
7. Hamilton, A. J., and Baulcombe, D. C. (1999) A species of small antisense RNA in posttranscriptional gene silencing in plants. *Science* 286, 950–2.
8. Pushparaj, P. N., and Melendez, A. J. (2006) Short interfering RNA (siRNA) as a novel therapeutic. *Clin Exp Pharmacol Physiol* 33, 504–10.
9. Pushparaj, P. N., Aarthi, J. J., Manikandan, J., and Kumar, S. D. (2008) siRNA, miRNA, and shRNA: in vivo applications. *J Dent Res* 87, 992–1003.
10. Barczyk, M. M., Olsen, L. H., da Franca, P., Loos, B. G., Mustafa, K., Gullberg, D., and Bolstad, A. I. (2009) A role for alpha11beta1 integrin in the human periodontal ligament. *J Dent Res* 88, 621–6.
11. Liu, Y., Arvidson, K., Atzori, L., Sundqvist, K., Silva, B., Cotgreave, I., and Grafstrom, R. C. (1991) Development of low- and high-serum culture conditions for use of human oral fibroblasts in toxicity testing of dental materials. *J Dent Res* 70, 1068–73.
12. Mustafa, K., Silva Lopez, B., Hultenby, K., Wennerberg, A., and Arvidson, K. (1998) Attachment and proliferation of human oral fibroblasts to titanium surfaces blasted with TiO2 particles. A scanning electron microscopic and histomorphometric analysis. *Clin Oral Implants Res* 9, 195–207.
13. Velling, T., Kusche-Gullberg, M., Sejersen, T., and Gullberg, D. (1999) cDNA cloning and chromosomal localization of human alpha(11) integrin. A collagen-binding, I domain-containing, beta(1)-associated integrin alpha-chain present in muscle tissues. *J Biol Chem* 274, 25735–42.

Chapter 7

Organ Cultures and Kidney-Capsule Grafting of Tooth Germs

Keishi Otsu, Naoki Fujiwara, and Hidemitsu Harada

Abstract

The study of organogenesis allows investigation of a variety of basic biological processes in the context of the intact organ. The ability to analyze teeth ex vivo during development has emerged as a powerful tool to understand how teeth are constructed and the signaling pathways that regulate these developmental processes. Here, we describe in detail our protocols for organ culture and kidney-capsule grafting of mouse tooth germs. These techniques allow us to reproduce the developmental process of tooth germs and estimate the effect of specific genes ex vivo, as well as are a tool for studies on the mechanisms of normal and abnormal tooth morphogenesis. They may also be applied to studies on other aspects of developmental biology and regenerative medicine.

Key words: Tooth development, Organ culture, Kidney capsule, Real-time imaging, Trowell method, Morphogenesis

1. Introduction

Tooth development proceeds with sequential and reciprocal epithelial–mesenchymal interactions, which are regulated by a number of soluble proteins (1–3). To date, most of the studies on tooth development have been done with conventional histological methods and cell culture. By those methods, however, it is very hard to analyze morphological changes and molecular dynamics in real time during tooth development. The major disadvantage of working with the developing tooth in vivo is the inaccessibility to observation while it develops within the alveolar bone.

Organ culture of tooth germs allows us to overcome this problem to a great extent. The development of cultured tooth germs remains comparable to the in vivo tooth development in terms of

Chrissa Kioussi (ed.), *Odontogenesis: Methods and Protocols*, Methods in Molecular Biology, vol. 887,
DOI 10.1007/978-1-61779-860-3_7, © Springer Science+Business Media, LLC 2012

both structure and function, which attributes to making them more suitable than cell culture for physiological studies. Moreover, the role of specific growth factors and genes under their influence can be assessed by using microbeads and/or microinjection (4, 5). In addition, we have established a new organ culture system for real-time imaging, which opens the door to the direct spatiotemporal analysis of the developing tooth.

Kidney-capsule grafting provides tooth germs with an in vivo biological environment. Under the kidney capsule, transplanted tooth germs retain their physiological features and the morphogenesis proceeds normally. A representative application of this method is for the examination of tooth germs from genetically modified mice. Such mice die during embryogenesis or shortly after birth (e.g., *Fgf10*$^{-/-}$ mice die immediately after birth), which makes it impossible to observe tooth development at later stages. Using kidney-capsule grafting, we can solve such problems and observe mouse tooth development for long periods of time (4).

Together, these two techniques provide information on the growth, differentiation, and development of tooth germs and the influences of various factors on these processes. The results obtained by use of these methods usually provide insight into the in vivo events; and thus, organ culture and kidney-capsule grafting can often replace conventional histological methods.

In this chapter, we describe our protocols for organ culture with real-time imaging and transplantation of mouse tooth germs under the kidney capsule, as well as give some technical tips.

2. Materials

All materials should be sterile.

2.1. Tooth Germ Extraction and Organ Culture

1. Scissors.
2. Fine forceps (Dumont, #5).
3. Culture dishes (35- and 100-mm dishes; 6-well plates).
4. Disposable plastic syringes (1 cm^3).
5. Needles (18 and 25 G).
6. Filters (0.1-μm pore size): The filters are cut into approx. 3 × 3-mm pieces and stored in 70% ethanol.
7. Metal grid (see Note 1).
8. Stereomicroscope with culture chamber (see Fig. 1).

2.2. Media

1. DMEM/F12.
2. Fetal bovine serum.

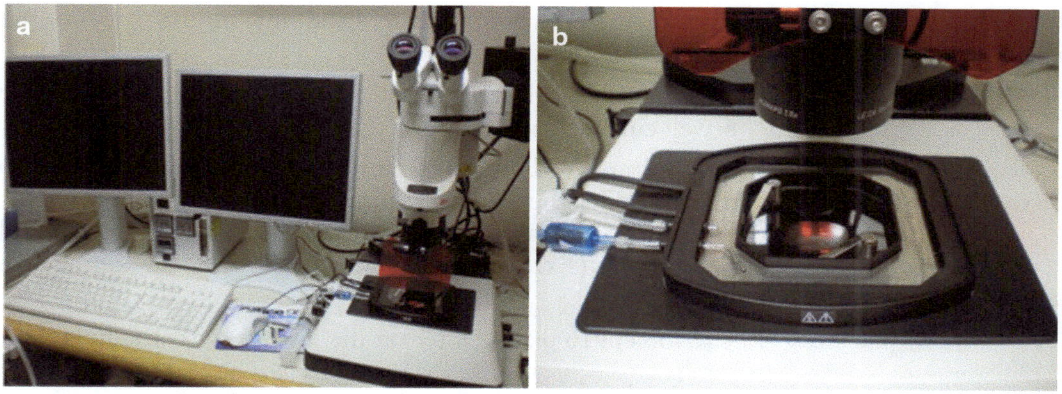

Fig. 1. Stereomicroscope with real-time imaging system for organ cultures. (**a**) Overall view of the system. (**b**) Magnified image of culture chamber on the stage. A culture dish is placed in the middle of the chamber.

3. Ascorbic acid.
4. Penicillin/streptomycin.
5. Phosphate-buffered saline (PBS), pH 7.4.
6. Working medium: DMEM/F12 supplemented with 50 U/ml penicillin/streptomycin.
7. Culture medium: DMEM/F12 supplemented with 10% fetal bovine serum, 50 U/ml penicillin/streptomycin, and 100 μg/ml ascorbic acid.

2.3. Kidney-Capsule Grafting

1. Glass Pasteur pipettes (see Note 2).
2. Scissors.
3. Forceps (Dumont, #5).
4. 25-G needles.
5. Disposable plastic syringes (1 cm^3).
6. Suture with needle (Syneture, 5-0 silk, C-13 19 mm needle).

2.4. General Equipment

1. Humidified incubator at 37°C and 5% CO_2.
2. Tissue culture hood.
3. 37°C water bath.
4. Stereomicroscope.

3. Methods

All procedures must be performed under sterile techniques with great attention. Use clean and detergent-free glassware.

3.1. Preparation of Culture Dishes for Organ Culture

1. Take filters stored in 70% ethanol and rinse in PBS in a 100-mm culture dish.
2. Use forceps to place metal grids in 35-mm culture dishes or 6-well culture plates (see Fig. 2).
3. Add culture medium, taking care not to allow air bubbles to remain under the grid. The surface of the medium should be at the same level as the top of the grid; i.e., the medium should not cover it. (Excess medium causes floating of the tooth germ on the filter).
4. Use forceps to place a filter on the grid to cover the hole.

3.2. Tooth Germ Extraction

All of the following steps should be performed on ice or at 4°C.

3.2.1. Molar Tooth Germ Extraction from Mouse Embryo

1. Rinse surgical equipment in 70% ethanol prior to use.
2. Sacrifice pregnant mice; remove embryos (E11–14) and place them in 100-mm dishes containing working medium on ice.
3. Wash several times in ice-cold working medium.
4. Use forceps and scissors to remove the embryo heads.
5. Remove lower jaw while observing under a stereomicroscope.
6. Use needles to remove tooth germs of molars from jaws.
7. Use needles to remove excess non-dental tissue.
8. Transfer tooth germs to fresh working medium and keep them on ice.

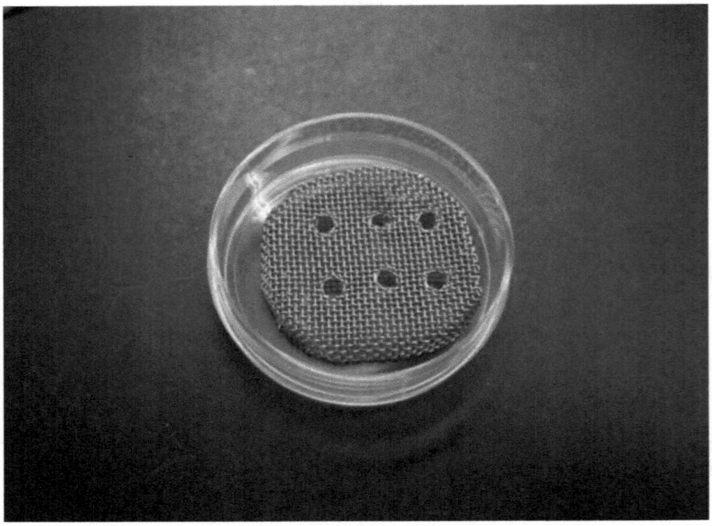

Fig. 2. The organ culture dish used for the Trowell method. The metal grid is placed in a 35-mm culture dish. The sample is transferred onto the filter, which is placed on the *holds of the grid.*

3.2.2. Incisor Tooth Extraction from Neonatal Mouse (see Note 3)

1. Wipe neonatal mice with 70% ethanol.
2. Sacrifice neonatal mice and remove heads under sterile conditions.
3. Transfer heads to ice-cold working medium in a 50-ml tube.
4. Wash several times with ice-cold medium.
5. Transfer head to a 100-mm dish containing working medium.
6. Peel off the skin from head.
7. Remove the lower jaw.
8. Using the tip of an 18-G needle or fine forceps, remove tooth germs under a stereomicroscope (see Note 4).
9. Eliminate superfluous tissues around the tooth germs (see Note 5) (Fig. 3).
10. Transfer the tooth germ to fresh working medium and keep on ice.

3.3. Organ Culture and Real-Time Imaging

1. Use forceps or glass pipettes to transfer tooth germs onto filters in culture dishes.
2. Check the position of the tooth germs; if needed, they can be rearranged by moving the filter.
3. Place the dish in a cell culture chamber attached to the stereomicroscope and incubate at 37°C, 5% CO_2.
4. Carry out microinjection of the desired substance (e.g., dye, plasmid, or siRNA). Alternatively, transfer agarose or heparin-coated acrylic beads bearing growth factors or inhibitors onto the tissue (4–6).

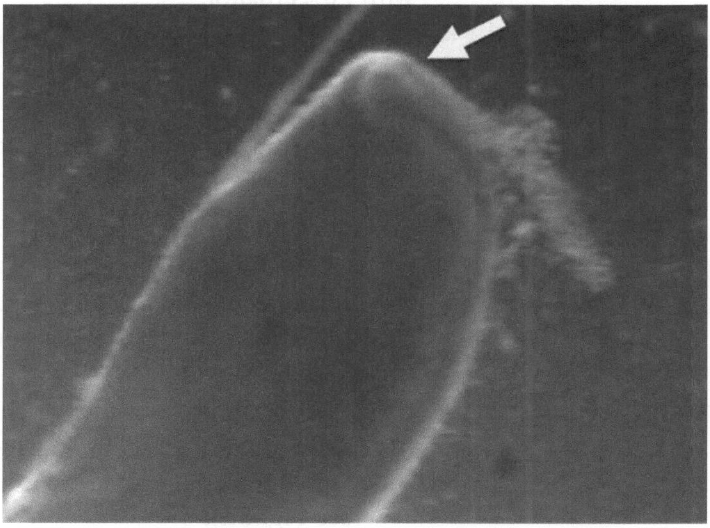

Fig. 3. Appearance of the mouse incisor apical end. *Arrow* indicates the apical bud.

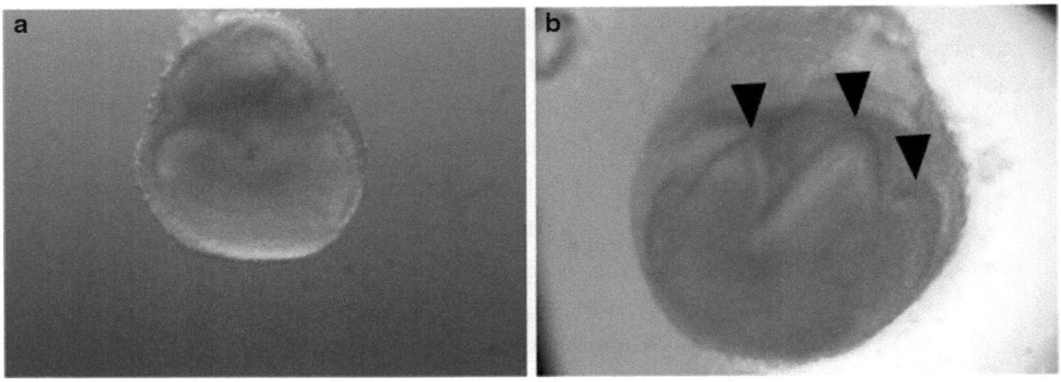

Fig. 4. Development of molar tooth germ in organ cultures. (**a**) The appearance of the dissected first molar tooth germ at E18, as seen under a stereomicroscope. (**b**) After 1 week in culture, cusp formation is clearly observed (*arrowheads*).

5. Take pictures at 30-min intervals to observe tissues (see Fig. 4).
6. After finishing real-time imaging, tissues may be used for subsequent experiments (e.g., histological analysis, gene analysis, protein analysis).

3.4. Kidney-Capsule Grafting

1. Use isoflurane inhalation to induce general anesthesia in male mice of the same strain or immunodeficient male mice (8–12 weeks old).
2. When mouse has been fully anesthetized, shave its left flank (see Note 6).
3. Swab the skin with povidone iodine and wipe the surgery site with 70% ethanol.
4. Locate the left kidney (just right of the spleen). Make a small incision in the skin and muscle and expose the peritoneum (see Note 7).
5. Make a small incision in the peritoneum and expose kidney (see Fig. 5a).
   (a) Lift kidney capsule and make a small breach with a 25-G needle (see Note 8).
   (b) Introduce the tip of a glass pipette through the breach and slide the tip under the capsule to carefully make a small pocket (see Fig. 5b).
   (c) Place the sample near the breach of the kidney (see Fig. 5c).
   (d) Pick up the capsule at the breach and carefully slip the sample into the pocket by using the tip of a glass pipette (see Fig. 5d).
   (e) Replace kidney in the abdominal cavity.

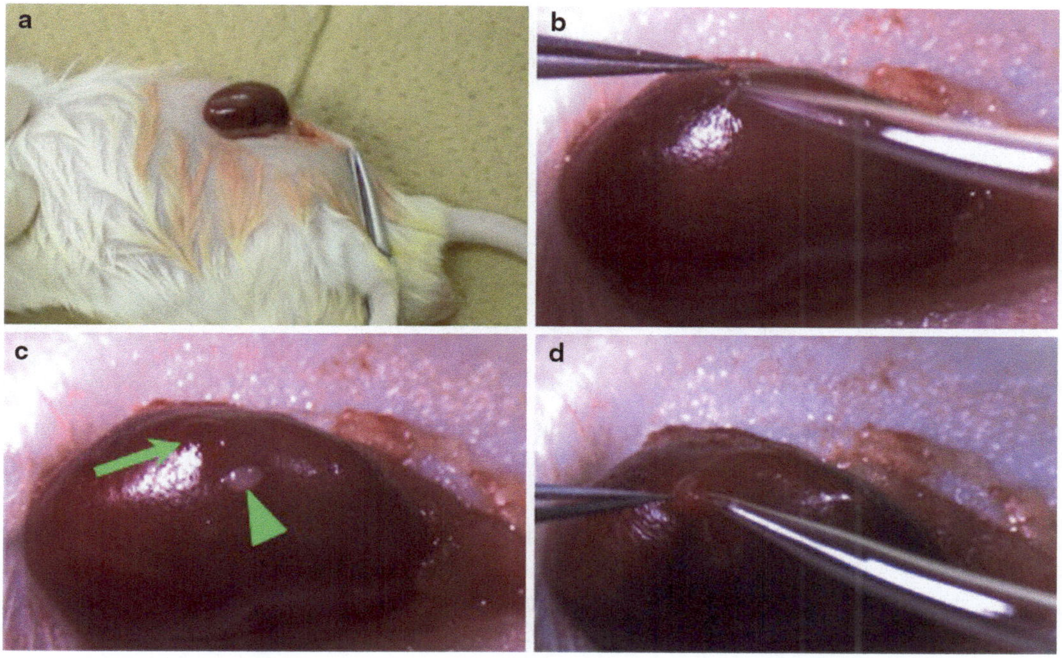

Fig. 5. Procedure for kidney-capsule grafting. (**a**) The left kidney is exposed and lifted from the abdominal cavity. (**b**) A small pocket is made under the kidney capsule by using the tip of a glass pipette. (**c**) A sample (*arrowhead*) is placed near the breach (*arrow*). (**d**) The capsule of the breach is picked up, and the sample is slipped into the pocket with the aid of the tip of the glass pipette.

(f) Close the musculo-peritoneal layer with 5-0 silk sutures.

(g) Close the skin with 5-0 silk sutures.

(h) Wash the surgical site with a povidone iodine swab.

(i) Place the mouse on a heating pad in a cage and monitor its recovery.

4. Notes

1. Cut 15–25-mm disks from stainless steel mesh (corrosion resistant, size of mesh 0.7 mm) and bend the edges to give a 2–3-mm height. Use a punch or a nail to make holes in the grid. These holes *contribute to a better* observation of the samples (see Fig. 2).
2. They are used for slipping the sample under the kidney capsule. To prevent gouging of the kidney or puncturing the capsule, the tips are closed to form a ball by heating them over a gas burner (see Fig. 6).

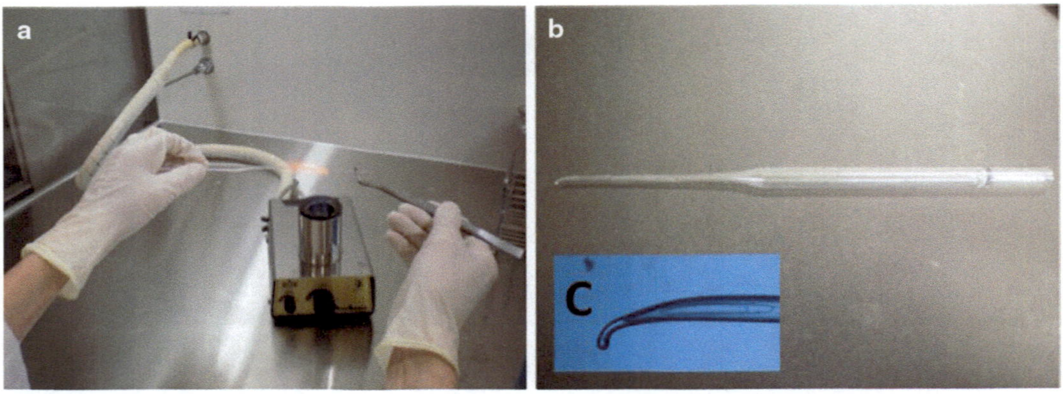

Fig. 6. Glass pipette for kidney-capsule grafting. (**a**) A glass pipette is pulled with heating in order to close the tip. (**b**) A glass pipette ready for use. (**c**) Magnified image of the tip.

3. Mouse incisors of postnatal day 1–2 are the most appropriate for organ cultures. Before postnatal day 7 is also suitable for extraction of tooth germs, since bone calcification is not complete at this stage. Forceps and needles can easily remove bone and other tissues.
4. Be very careful not to separate the epithelial layer (ameloblasts) and apical bud from the mesenchymal layer (odontoblasts).
5. Removing excess of non-dental tissue and dental follicle facilitates the observation of the epithelial layer and apical bud.
6. Only the left kidney is used for transplantation. Right kidney stays intact.
7. A small incision helps to keep kidney raised and exposed.
8. Prevent gouging of the kidney and the attendant bleeding; otherwise, it becomes difficult to recognize the opening of the breach.

Acknowledgments

This work was supported, in part, by KAKENHI (20890208 to KO) and Iwate Medical University, School of Dentistry Open Research Project (2007-2011) from the Ministry of Education, Culture, Sports, Science, and Technology of Japan.

References

1. Thesleff, I., and Mikkola, M. (2002) The role of growth factors in tooth development, *Int Rev Cytol 217*, 93–135.
2. Thesleff, I., Keranen, S., and Jernvall, J. (2001) Enamel knots as signaling centers linking tooth morphogenesis and odontoblast differentiation, *Adv Dent Res 15*, 14–18.
3. Jernvall, J., and Thesleff, I. (2000) Reiterative signaling and patterning during mammalian tooth morphogenesis, *Mech Dev 92*, 19–29.

4. Yokohama-Tamaki, T., Ohshima, H., Fujiwara, N., Takada, Y., Ichimori, Y., Wakisaka, S., Ohuchi, H., and Harada, H. (2006) Cessation of Fgf10 signaling, resulting in a defective dental epithelial stem cell compartment, leads to the transition from crown to root formation, *Development 133*, 1359–1366.
5. Harada, H., Kettunen, P., Jung, H. S., Mustonen, T., Wang, Y. A., and Thesleff, I. (1999) Localization of putative stem cells in dental epithelium and their association with Notch and FGF signaling, *J Cell Biol 147*, 105–120.
6. Harada, H., Ichimori, Y., Yokohama-Tamaki, T., Ohshima, H., Kawano, S., Katsube, K., and Wakisaka, S. (2006) Stratum intermedium lineage diverges from ameloblast lineage via Notch signaling, *Biochem Biophys Res Commun 340*, 611–616.

Chapter 8

Evaluation of Skull and Tooth Morphology and Mineralization Using High-Resolution X-Ray Tomography

Brian K. Bay

Abstract

High-resolution X-ray tomography (microCT) is increasingly available in research settings, and is a valuable tool in the study of mineralized tissue development. At resolutions of 2–20 μm, achievable for typical *murine* scale samples, it provides nondestructive visualization of three-dimensional tissue morphology and a limited ability for quantitative measurement of developmental parameters. Sample preparation is simple and can be tailored for compatibility with other biological assays. Here, we describe the application of microCT to the investigation of lower incisor development in the context of overall skull morphology.

Key words: Microtomography, Skeletal morphology, Tooth morphology, Mineralization

1. Introduction

In studying the developmental biology of mineralized tissues and structures, quantitative data on three-dimensional morphology, material distribution, and material quality expressed in experimental systems is of critical importance. This is particularly true for studies of odontogenesis, as tooth shape is a key aspect of proper functional mechanics, and carefully tuned material characteristics are essential for long-term durability.

With mineralized tissues, the serial sectioning approaches generally employed for histological evaluation are compromised. Microtome cutting after paraffin embedding is only possible after demineralization, and this eliminates the ability to evaluate mineral characteristics. Thin-blade sectioning and grinding after rigid plastic embedding preserves mineralization details, but it is a time-consuming procedure and destroys considerable sample material.

Chrissa Kioussi (ed.), *Odontogenesis: Methods and Protocols*, Methods in Molecular Biology, vol. 887,
DOI 10.1007/978-1-61779-860-3_8,

The challenge is especially acute when *murine* models are employed. Length of the mature incisors is in the millimeter range, and cross-sectional dimensions are on the order of a tenth the length. To map three-dimensional mineral distribution over this spatial scale is impractical with any direct-sectioning approach.

An effective alternative is virtual sectioning using laboratory X-ray tomography (microCT). Since the advent of clinical X-ray tomography (CT or CAT scanning), it has been clear that many laboratory investigations would benefit from similar technology. Relaxing the limitations on X-ray dose and scan time imposed when live humans are the subjects offers greatly improved spatial and attenuation resolution. Impediments to microCT application have largely been a matter of cost, access to imaging equipment, and computer capacity for information processing, but these barriers have gradually eroded.

It is currently possible to generate highly detailed laboratory-based microCT images of teeth at the human scale with geometry, mineral distribution, and mineral characteristics accurately quantified (1–4). Synchrotron-based X-ray tomography offers additional capabilities (5), but is difficult to incorporate into routine sample evaluation and is available on a limited basis. The *murine* scale remains a challenge due to the small size of the samples, but useful information can be derived from conventional microCT equipment and new generations of scanning systems are offering greatly enhanced capabilities (6).

The focus here is on using standard, commercially available microCT equipment as an adjunct to more detailed histological methods for *murine* teeth evaluated within the context of overall skeletal development. Scanners of the cone-beam architecture are featured (7), but most of the basic principles will apply to other architectures as well. The objectives are to evaluate bone tissue development, establish incisor morphology and size, and verify the existence of mineralization in areas of key cellular activity, with the overall goal of comparing wild-type and genetically manipulated strains.

2. Materials

Preparation of biological samples for microCT is relatively simple. At the most basic level, nothing is required other than a container of a size compatible with the imaging hardware, and capable of holding the samples *still* during the scanning process. Results are greatly improved, however, with basic knowledge of the scanning process and how X-rays interact with samples, containers, and storage media.

2.1. MicroCT Scanner

1. The microCT images shown here were produced by a Scanco μCT 40 (SCANCO Medical, Brüttisellen, Switzerland). There are other manufacturers of laboratory microCT equipment. It is an active area of equipment development, and new manufacturers and models appear regularly.
2. The most basic scanner characteristics involve sample size, resolution, and useable X-ray energies. Most researchers are interested in the ability to discriminate features of a given size within samples of a particular material system. Many factors affect the ability to achieve a specific imaging goal, and it is not within the scope of this document to be comprehensive, but some general guidelines can be stated.
3. Perhaps the most important aspect of tomographic imaging is the relationship between sample size and spatial resolution. There are limits to the resolution that can be achieved for a given sample size, and as resolution demands *increase*, sample size *decreases* (see Note 1). Local tomography (8) relaxes this relationship somewhat, and the specific size/resolution relationship will vary among imaging systems, but it is always present.
4. There is also an absolute limit to achievable resolution (9). This is generally around the focal spot size of the X-ray source, and does not go below a few microns for most laboratory systems (see Note 2).
5. Routine scans of centimeter-scale samples at tens to hundreds of micron resolution can be completed within an hour. Expect scan times of many hours for small-sample, high-resolution work (see Note 3).
6. A computer-based tomographic reconstruction process follows sample scanning. This produces a voxel volume analogous to the pixel area of simple digital images. The voxel volume is the output of the process, and is the raw data for subsequent image processing (see Note 4).
7. Mineralized tissues are imaged in the range of 40–80 kVp, a common level of energy for laboratory X-ray sources. Specialized sources are required for energies below about 20 kVp, the useful range for un-mineralized tissues.
8. Just as important as the ability to generate suitable image data is the ability to process it (see Note 5). All manufacturers offer some level of image-processing software support, often very specialized for particular scanning needs. Third-party software for processing of volume image data is also available. Expect a relatively steep learning curve, or tapping local expertise, if custom software for novel measurement tasks is required.

9. Fortunately, the need for quantitative imaging of the mouse skeleton in support of biomedical research is long-standing and widespread (10, 11), and manufacturers have focused effort on both hardware and software tailored for that task. If you are fortunate enough to purchase a scanner, then work with a reputable manufacturer to specify a system (scanner model, modifications, accessories, software) suited to your specific needs. If forced to work with a less than ideal but available system, then be prepared to accept some limitations in capabilities and invest resources into materials and procedures to get the best results possible.

2.2. Sample Container

1. Select a container that just accommodates the sample, with minimal extra room, and also fits within the imaging equipment (see Note 6). If specific containers are supplied with the scanning equipment, use those and select the most appropriate size.
2. Select a container material that has low X-ray attenuation characteristics with respect to the sample (see Note 7). Plastic containers are generally suitable for use with mineralized samples, but can contain additives or surface markings with increased attenuation. Glass containers may be too attenuating.

2.3. Sample Preparation and Environment

1. There are no specific sample preparation requirements for microCT; that is one of the chief advantages of the method.
2. It is not required to immerse the sample for microCT imaging, but it is often desirable from both sample preservation and image quality perspectives (see Note 8).
3. If a sample is destined for subsequent histological or biochemical processing, then requirements for that methodology should dominate the selection of an immersion fluid.
4. If there are no other overarching requirements, a phosphate-buffered saline (PBS) solution is adequate for immersion of mineralized tissue samples.
5. A sample may be scanned frozen, as long as it does not thaw out and move during the scan. There is an increasing interest in cryostages among microCT users with biological applications, but unfortunately the temperature levels required for effective mRNA preservation are not generally available.
6. Focal, high-atomic number inclusions (metal rods, screws, beads, etc.) should be excluded from the sample, if possible. These will generate reconstruction artifacts that may interfere with image evaluation (see Note 9).

3. Methods

1. Position the sample in the container and, if necessary, wedge in place with foam or plastic to prevent motion (see Note 10).
2. Orient all samples in a consistent manner within the sample container (see Note 11), and if a key feature (such as an incisor) is elongated in shape, orient the long axis of the feature parallel to the scanner rotation axis (see Note 12).
3. Verify with scout or orientation views that sample container and any sample-positioning devices do not interfere with or dominate X-ray attenuation.
4. Fill the sample container with an appropriate immersion fluid, ensuring that air bubbles are eliminated to the greatest extent possible. If bubbles within the sample create difficulties with subsequent processing of the image data, hold the immersed sample under vacuum until bubble elution ceases.
5. Place the sample within the scanner, and follow manufacturer's recommendations for establishing X-ray exposure settings to the extent they are available. Bracket the recommended X-ray parameters and conduct a series of relatively low-resolution scans before final parameters are set (see Note 13).
6. The tomographic reconstruction process is fully automated in most commercial scanning systems. Take advantage of any options available for reducing the reconstruction volume by targeting a region of interest (see Note 14).
7. With the tomographic reconstructions completed, the data processing moves to the rendering and measurement stages. A basic step in producing renderings (visualizations) from microCT data is segmentation by voxel value (establishment of a threshold range). Voxel values above and below the set levels are excluded from the rendering process. It is relatively simple to segment mineralized from un-mineralized tissues, at least for the purpose of visualization, and the process is frequently conducted with interactive software tools.
8. These visualizations, especially when viewed in three dimensions and interactively, are often adequate to verify normal skeletal morphology (Fig. 1). In this wild-type sample, the skull bones appear normal and the incisors can be seen emerging from the lower mandible.
9. For quantitative measures of segmented volume, the threshold process is more critical, as small differences in the inclusion

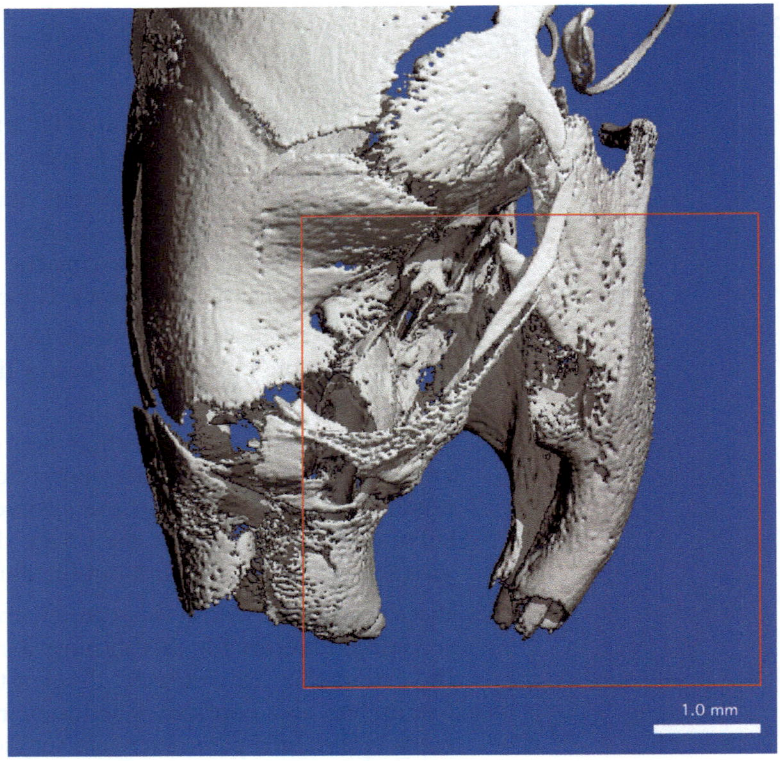

Fig. 1. Surface-shaded rendering of the entire sample with threshold levels set for mineralized tissues. The *red box* indicates the location of the subsequent cut-away rendering.

voxel range can have a large influence on measurement results. Accurate measurements require a calibration process (see Note 15).

10. Virtually cutting away portions of the sample and rendering the remainder reveal interior morphological details (Fig. 2). This step also establishes the orientation of interior features, such as the incisors within the lower mandible. Proper orientation is the key for measurements of linear dimensions.
11. A virtual cutting plane (see Note 16) through the lower mandible, and perpendicular to the long axis of the incisors, reveals tooth morphology and identifies key mineralization regions (Fig. 3).
12. Measurements of tooth cross-sectional dimensions precise enough for comparison of experimental groups are possible from properly oriented virtual cutting planes (tooth diameter is approximately 200 μm).
13. Indications of mineralization differences can be seem in the virtual cutting plane image (denser material on the flat labial side of the tooth, focal areas of mineralization within the tooth

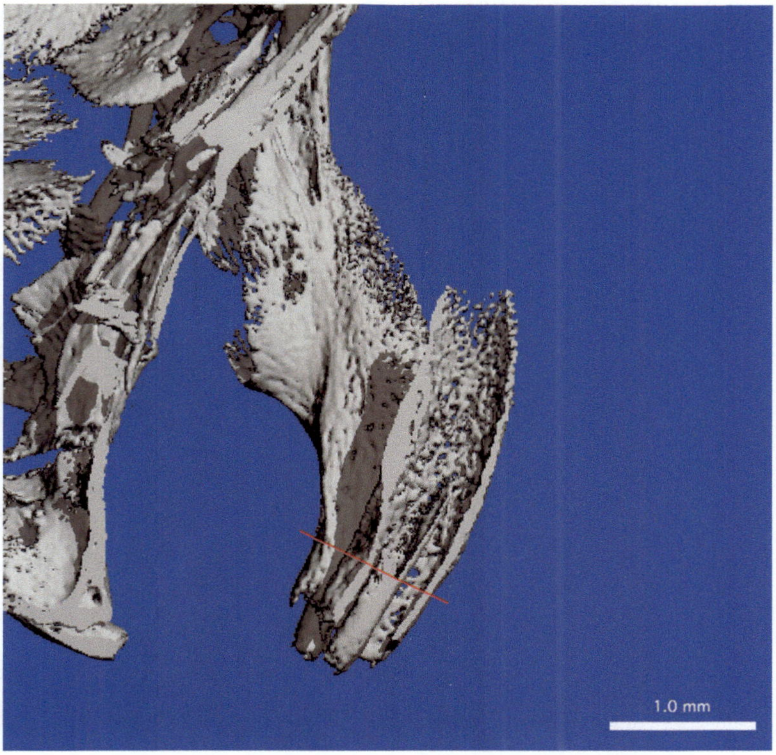

Fig. 2. Cut-away view of the skull showing the position and orientation of the incisors within the lower mandible. The *red line* indicates the location and orientation of the subsequent planar view.

envelope), but they cannot be quantified at this resolution. The thickness of the tooth wall is approximately 20–40 μm, and the reconstruction resolution is 10 μm, leaving only 2–4 voxels across the wall thickness.

14. There is no method to improve resolution through image processing. The visualization could be made more visually appealing by smoothing and contrast/brightness adjustments, but this would not improve measurement values.

15. Due to the sample size and resolution relationships imposed by the scanner, the only recourse is to physically cut a smaller sample (remove the lower mandible, for example) and re-scan at higher resolution within a smaller sample container. Resolution could be moved closer to the absolute limit of the scanning system in this manner, but for standard laboratory microCT this will not be better than 1–5 μm.

16. Resolution of details below a micron is beyond the reach of laboratory microCT based on standard X-ray tubes. If tomography at this level of detail is required, samples must be evaluated at a synchrotron X-ray source.

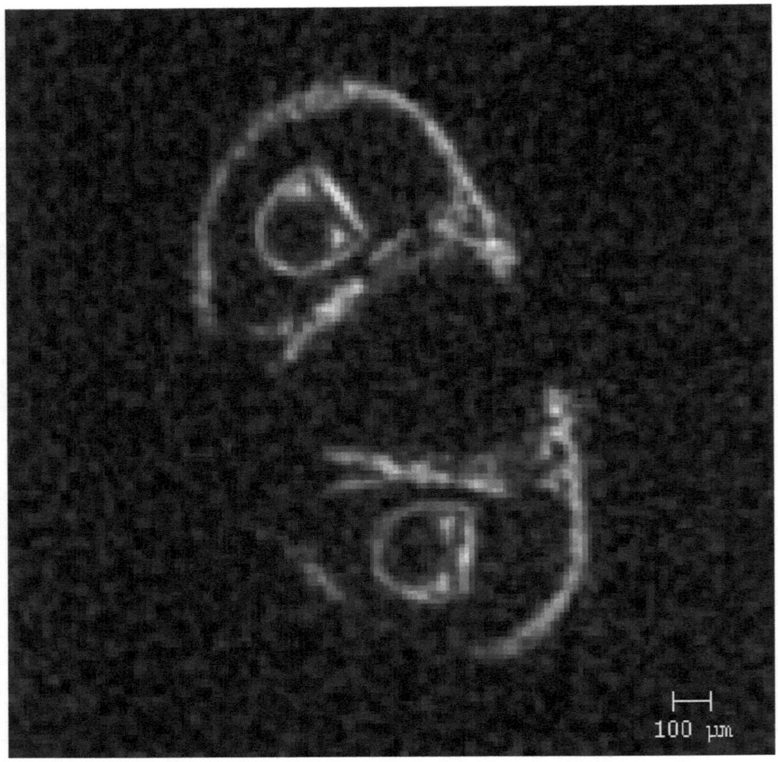

Fig. 3. An interpolated plane created perpendicular to the incisor axis.

4. Notes

1. For a well-designed scanning system, the resolution will be roughly the sample (including container) diameter divided by the reconstruction slice size. For example, if the sample container is 15 mm in diameter and the reconstructed slice is 1,024 × 1,024, an individual voxel represents approximately 15 µm. Voxel size is not equivalent to resolution, but it gives an indication of resolvable feature size. It is possible to over-reconstruct the image volume, and generate voxels much smaller than any detectable feature, but a well-designed scanning system will not do this.
2. Cone-beam tomography involves collecting a long series of shadow images, perhaps a thousand in total, as the sample is rotated within the X-ray beam. Sharpness of the shadow images (size of the penumbra) is a key factor in the ultimate resolution of the reconstructed images, and is related to the focal spot size of the X-ray source.
3. For common laboratory X-ray tubes, the beam is generated by firing a stream of electrons (the tube current) at a small region

(the focal spot) of a metal target. The focal spot must be kept very small to maintain sharpness of the X-ray images, but this limits the tube current to very low levels to prevent melting of the target. Noise in the images is related to the total X-ray flux out of the tube, and therefore low-noise images require long scan times.

4. The voxel volume is often stored as a series of digital images that appear as virtual cross-sectional slices of the sample, perpendicular to the sample rotation axis. They are not images per se but maps of X-ray attenuation capability for a finite-thickness slab of material at that location along the sample rotation axis. Taken together, they map internal X-ray attenuation capability for the entire sample volume.
5. Contemporary software for manipulation of tomographic scan data largely abandons the slice-based storage of the initial voxel *reconstruction* and treats the entire volume as a three-dimensional data set. The data becomes a source for *renderings* of many kinds: slices along planes obliquely oriented within the volume, extraction of surfaces falling within an attenuation range, transparency- and shading-enhanced visualizations of interior structures, etc. The renderings facilitate *measurements* that quantify lengths, areas, and volumes of interior structures. The process of going from raw scan data to specific measurements in a simple and efficient manner, although vastly better than when scanning hardware first appeared, remains an active area of development. Expect considerable time investment for complex measurement goals.
6. Check dimensions carefully; scanners can be very limiting with respect to sample size. Resolution of the scan is also influenced by position of the sample with respect to the X-ray source. Extra space around the sample is a detriment to optimal imaging since both sample and container must fit fully within the useable X-ray beam and will limit the amount of reconstructed voxel space that covers the actual sample.
7. For imaging dominated by continuous energy or Bremsstrahlung X-rays, the primary factors influencing X-ray attenuation are atomic number and density. Materials with even small amounts of a high-atomic-number component can be quite attenuating as are high-density materials of low-atomic-number (e.g., tooth enamel). Be cautious even with marker labeling of a sample container. Inks often contain high-atomic-number components and can remain visible during X-ray imaging.
8. Immersion of samples in a fluid medium often improves tomographic reconstructions due to the slight X-ray beam filtering effect and the consequent reduction in beam hardening artifact. The sample should not float, however, as this will virtually guarantee motion during the scan.

9. Metal particles are a common source of microCT imaging artifact. The tomography computer process requires the calculation of X-ray attenuation (ratio of transmitted to incident beam intensity), and if the X-ray beam is completely blocked, this information is lacking. The result is "star" artifacts, bright bursts within the image data that obscure information in adjacent areas.
10. The sample must remain absolutely still within the container during scanning. Any motion will produce significant imaging artifact. Scan times range from minutes to hours, and tissue samples can exhibit viscoelastic creep. Keep the sample fixed in place without undue pressure.
11. From a theoretical perspective, it does not matter how a sample is oriented since the imaging captures the entire sample volume. But from a practical perspective, it greatly simplifies post-processing of image data if the samples are consistently oriented.
12. The tomographic reconstruction process generates an *estimate* of X-ray attenuation variation within the sample volume. The estimation process *assumes* no X-ray scatter and uniform beam energy. The assumptions are a better match to reality when the X-ray beam traverses the cross section of elongated objects instead of the long axis.
13. How well features of interest are resolved is not fully apparent until the tomographic reconstruction is examined. Trial reconstructions at reduced resolution, and confirmation at full resolution, is the best approach before a full sample series is processed.
14. Even with contemporary computer capabilities, storage of tomographic reconstructions is cumbersome. If raw projection data is stored along with the reconstructed voxels, a relatively modest $1,024^3$ microCT scan will require between 2 and 4 GB of storage, depending on how the data is represented. For a $2,048^3$ volume, each scan requires 16–32 GB of storage. Optional data formats for fast volume manipulation can require storage of several times these volumes.
15. The best method of calibrating volume measurements from microCT data is to image a standard of known volume. The dimensions and material of the standard must be representative of the actual samples. The standard is imaged under identical conditions as the samples, and volume measurements are derived for a range of voxel threshold values. The values that generate the correct volume result are used for the actual samples.
16. Interpolation is used to create *virtual* cutting planes oriented at any angle within the voxel volume. This is a key advantage over direct sectioning techniques. It is difficult to precisely control the orientation of an *actual* section cut from the sample, and orientation errors will distort linear measurements. With virtual sectioning, the orientation is precisely controlled.

Acknowledgments

We are grateful for the assistance of Urszula Iwaniec, who conducted the microCT scanning appearing here. The microCT facility is part of the Bone Research Laboratory, Nutrition and Exercise Sciences, College of Health & Human Sciences, Oregon State University.

References

1. Seifert A., Flynn M.J., Montgomery K., and Brown P. (2004) Visualization of X-ray microtomography data for a human tooth atlas: *Proc SPIE* 5367, 747–757.
2. Rossi M., Casali F., Romani D., Bondioli L., Macchiarelli R., and Rook L. (2004) MicroCT Scan in Paleobiology: Application to the Study of Dental Tissues: *Nucl Inst and Meth in Phys Res B* 213, 747–750.
3. Davis G., Evershed A., Elliott J., and Mills D. (2010) Quantitative X-ray microtomography with a conventional source: *Proc. SPIE 7804, Developments in X-Ray Tomography VII.*
4. Dalstra M., Cattaneo P.M., Beckmann F., Sakima M.T., Lemor C., Laursen M.G. and Melsen B. (2006) Microtomography of the human tooth-alveolar bone complex: *Proc. SPIE* 6318, *Developments in X-Ray Tomography V.*
5. Zaslansky P., Zabler S., and Fratzl P. (2010) 3D variations in human crown dentin tubule orientation: A phase-contrast microtomography study: *Dental Materials* 26, 1–10.
6. Feser M., Gelb J., Chang H., Cui H., Duewer F., Lau S.H., Tkachuk A., and Yun W. (2008) Sub-micron resolution CT for failure analysis and process development: *Meas Sci and Tech* 19.
7. Feldkamp I.A., Davis L.C., and Kres J.W. (1984) Practical Cone-beam Algorithm: *J Opt Soc Am* 1, 612–619.
8. Katsevich A. (2006) Improved cone beam local tomography: Inverse Problems 22, 627–643.
9. Wang Y., Duewer F., Kamath S., Scott D., Yun W. (2004) A novel X-ray microtomography system with high resolution and throughput: *NSTI Nanotech 2004* 3, 503–507.
10. Sasov A. (2001) High-resolution in-vivo micro-CT scanner for small animals: *Proc. SPIE* 4320, 705–710.
11. Latson L., Kuban B., Bryan J., Stredney D., Davros W., Midura R., Apte S., and Powell K. (2003) X-ray Micro-Computed Tomography System: Novel Applications in Bone Imaging: *Annual International Conference of the IEEE Engineering in Medicine and Biology – Proceedings* 2, 1058–1061.

Chapter 9

Electron Microscopy

Hans U. Luder and Margrit Amstad-Jossi

Abstract

Correlative light (LM) and transmission electron microscopic (TEM) analysis is useful, if ultrastructural details of cells need to be related to functional aspects which can only be examined at the LM level. The first protocol presented here introduces a relatively simple way of obtaining TEM images which, on the one hand, reveal ultrastructural details of individual cells and, on the other hand, are large enough to allow a correlation with light micrographs. The second protocol describes a technique for estimating mineral densities of hard tissues using backscattered electron images obtained with a scanning electron microscope. This technique can be used to analyze the mineralization processes which occur throughout tooth formation.

Key words: Correlative transmission electron microscopy, Scanning electron microscopy, Calibrated backscattered electron imaging, Hard tissues, Mineral density estimation

1. Introduction

Electron microscopy is a well-established technique comprising a wide range of methods for transmission electron microscopy (TEM) and scanning electron microscope (SEM) applications; therefore, numerous guidelines already exist which deal with both basic and advanced techniques (1–5). The question then is whether ultrastructural investigations related to odontogenesis require any special procedures. Although the protocols described below are applicable also in other fields of research, they are dealt with here because they are particularly useful for TEM and SEM investigations of tooth formation.

The first procedure provides a relatively simple way of obtaining correlative light microscopy (LM) and TEM data. A common problem faced when examining specimens with the TEM is the fact that its magnification range starts approximately where that of the

Chrissa Kioussi (ed.), *Odontogenesis: Methods and Protocols*, Methods in Molecular Biology, vol. 887,
DOI 10.1007/978-1-61779-860-3_9, © Springer Science+Business Media, LLC 2012

LM ends. As a result, it can be difficult to identify the area of interest, which has already been established at the LM level. This, however, is important when observing the subtle changes in morphology, which, for example, occur in the course of ameloblast and odontoblast differentiation and are associated with alterations in signaling which are usually investigated with the LM. The solution provided by the procedure described below uses a series of TEM micrographs, which are taken at relatively high magnifications of large ultrathin sections. These micrographs are then assembled into a composite image, which provides an overview of an area which can be easily correlated with conventional LM images and allows the recognition of ultrastructural details of individual cells without digital enlargement (Fig. 1). The method does not, however, necessarily yield LM and TEM views of identical cells, as is the case when following some of the techniques described in the guidelines published by Hayat (6).

The second procedure provides an estimate of the mineral density of hard tissues using micrographs obtained from flat, polished specimens with the backscattered electron (BSE) detector of an SEM (Fig. 2). This is useful because tooth formation is, to a great extent, a calcification process. In particular, the initial mineralization and subsequent maturation of enamel result in mineral gradients, which can be analyzed using the described technique. It relies on the almost linear relationship between the intensity of the BSE signal and the mean atomic number of a compound. Hence, if assumptions can be made as to the chemical nature of the constituent parts of a tissue, their concentrations can be derived from the gray level of a BSE micrograph. Various attempts have been made to estimate mineral densities in this way, not only of bone (7–13), but also of sound and diseased teeth (14, 15). Unfortunately, the procedures applied have not been described systematically or in detail. The step-by-step guidelines below are an attempt to do that. If the assumptions underlying the calculation are taken into consideration and data are interpreted carefully, the obtained mineral density estimates provide more realistic information than a simple analysis of gray levels (16), particularly if data are derived from several areas of one specimen and can be analyzed in a comparative way (Fig. 3b).

2. Materials

Since the suitability and preparation of materials used for electron microscopic tissue processing, embedding, and sectioning are dealt with comprehensively in earlier work (1, 3–5, 17), the following list comprises only those solutions, materials, and equipment which we routinely use in our laboratory.

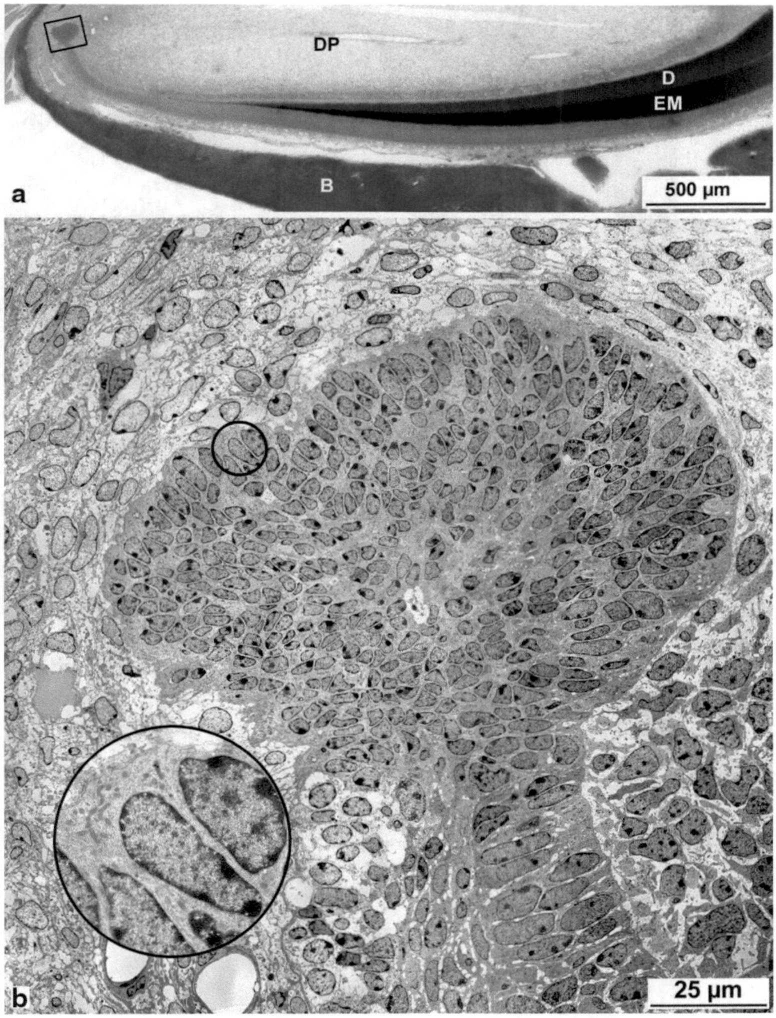

Fig. 1. Light (**a**) and transmission electron (**b**) micrographs of the cervical loop of a lower incisor from a 3-week-old mouse. The structure marked by the rectangle in (**a**), resembling the head-like structure described by Harada and Ohshima (19), is displayed in (**b**). The inset in (**b**) is a closer view, obtained without any digital enlargement, of the detail marked by the *circle*. *B* bone, *D* dentin, *DP* dental papilla, *EM* enamel matrix. Original magnifications (**a**) ×20, (**b**) ×1,950.

2.1. Specimen Processing

1. Half-strength Karnovsky's fixative: 2% paraformaldehyde, 2.5% glutaraldehyde, 0.025% $CaCl_2$, 0.02 M Na-cacodylate (pH 7.2–7.4).
2. Phosphate-buffered fixative: 4% paraformaldehyde, 0.2% glutaraldehyde, 0.1 M phosphate buffer (pH 7.2–7.4).
3. Postfixation: 1.33% Os-tetraoxide, 0.067 M s-collidine.
4. Wash buffer: 0.185 M Na-cacodylate (pH 7.2–7.4).
5. Decalcification: 10% EDTA (Titriplex III), 0.2% glutaraldehyde.

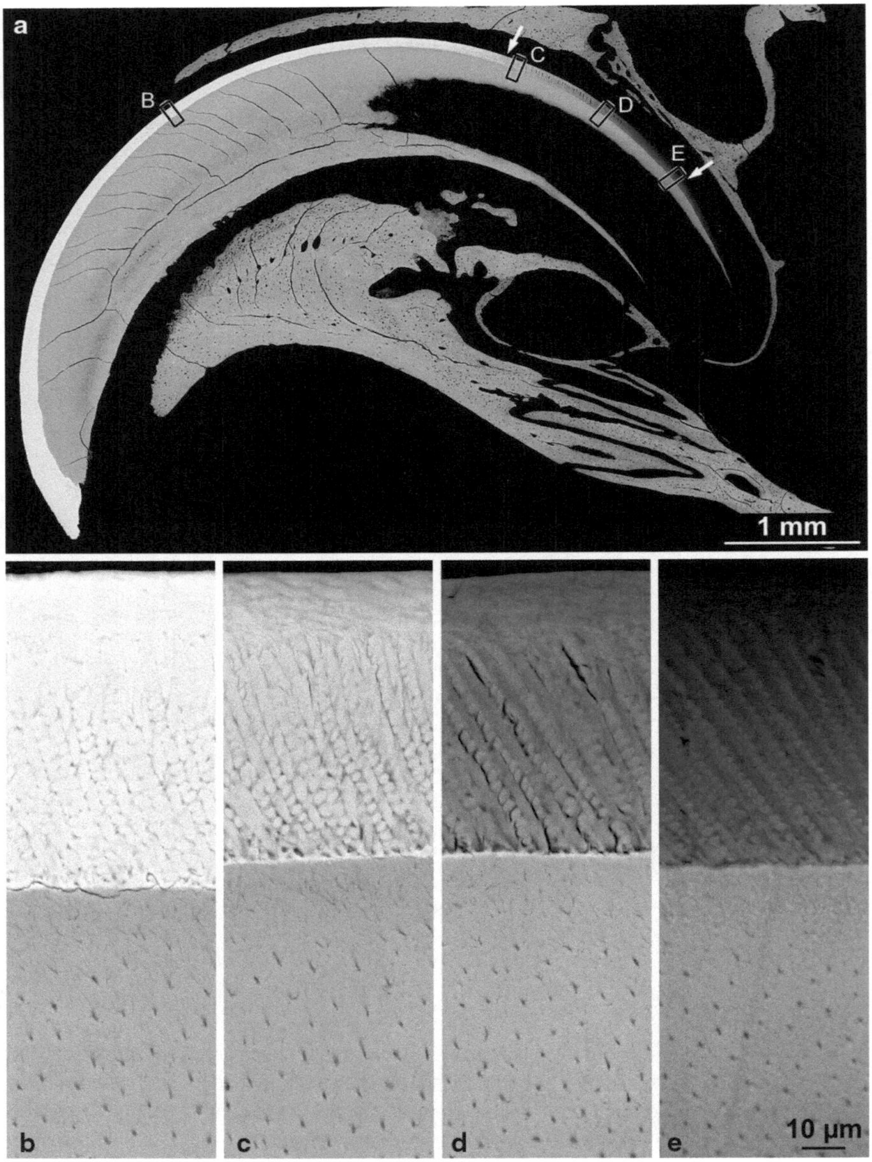

Fig. 2. Calibrated backscatter electron micrographs of an upper incisor from a 3-week-old mouse, which were used for estimating the mineral densities (MDE) represented graphically in Fig. 3. (**a**) Overview; the *two arrows* delineate the zone of enamel maturation which was analyzed as indicated in Fig. 3a; the rectangles labeled B, C, D, and E mark the location of the details shown in (**b–e**). (**b–e**) Details of the enamel and peripheral dentin after eruption (**b**) as well as at the end (**c**), in the middle (**d**), and at the beginning (**e**) of the zone of enamel maturation. MDEs across the enamel in these areas are graphically displayed in Fig. 3b. Original magnifications (**a**) ×100, (**b–e**) ×2,000.

2.2. Embedding

1. Epon: 122 g Epon 812, 79.5 g DDSA, 53.5 g MNA, 3.82 gN, *N*-benzyldimethylamine.
2. Beem capsules: Size 00 with 8 mm inner diameter.
3. Technovit 7200 VLC.

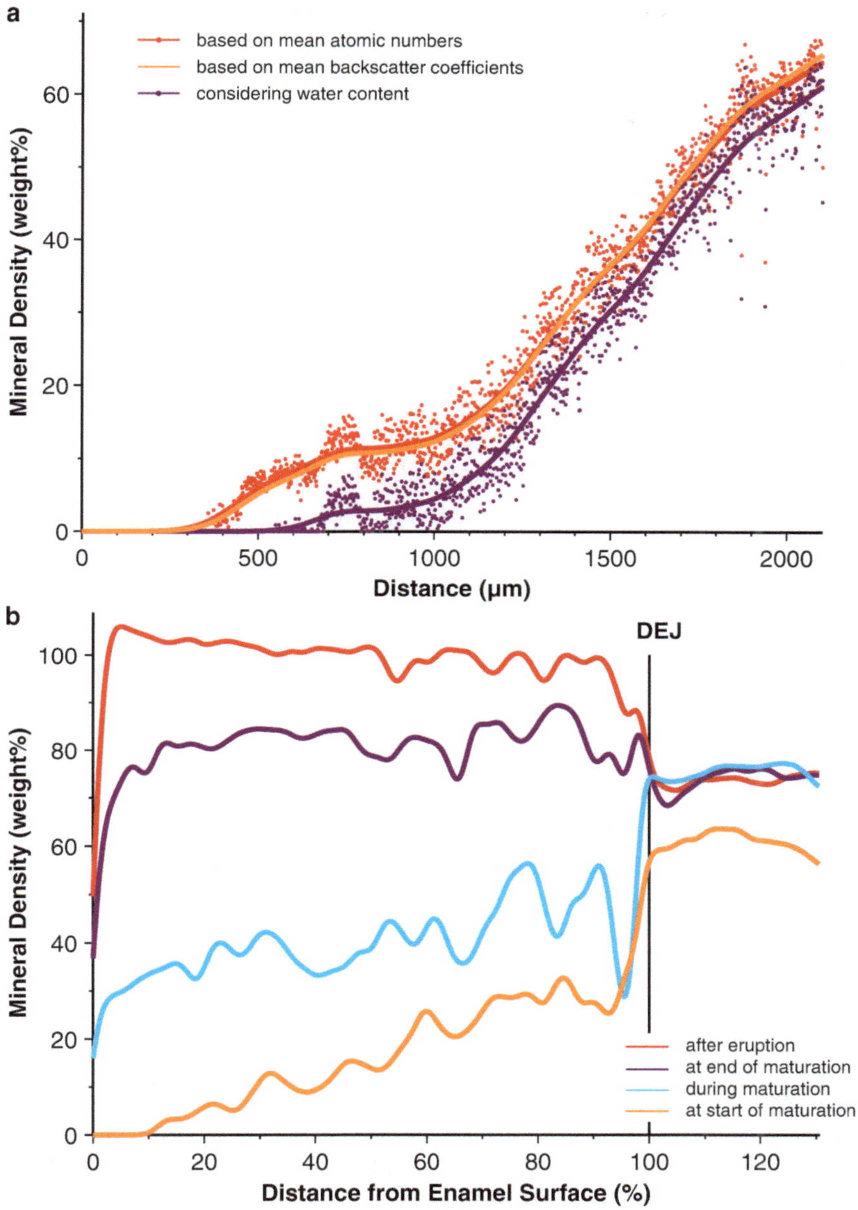

Fig. 3. Graphical representations of mineral density estimates (MDEs) made on the micrographs shown in Fig. 2. (**a**) MDE obtained from a five pixel-wide line in the middle of the enamel layer from the beginning to the end of enamel maturation indicated by the *arrows* in Fig. 2a. MDEs plotted as *solid line* and *dots* were derived on the basis of mean atomic numbers, MDEs represented as irregularly *dotted line* (close by the *solid line*) were obtained on the basis of mean backscatter coefficients, and MDEs plotted as *open circles* and the regularly *dotted line* were calculated under the assumption that water was present in a proportion of 4 parts water per 1 part organic matrix. (**b**) MDEs obtained from five pixel-wide lines across the areas of enamel and peripheral dentin shown in Fig. 2**b–e**. It was assumed that no water was present either in enamel or dentin. The distance from the surface is indicated as percent of the total enamel thickness (0–100%) and of the total dentin thickness (100–200%); as a consequence, the dentin–enamel junction (DEJ) is located at a distance of 100% from the surface.

4. Technovit 7200 embedding molds.
5. EXAKT light polymerization unit.

2.3. Sectioning

1. Diamond knives: Histo diamond knife with 6-mm cutting edge or ultra diamond knife with 2.5-mm cutting edge.
2. Grids: Parallel bar copper grids coated with a collodion support film (made with 2% collodion in amyl acetate) with about 5-nm-thick carbon coat.

2.4. Staining and Contrasting

1. Toluidine blue: 1% Toluidine blue O, 1% Borax.
2. Uranyl acetate: Saturated solution in double-distilled water.
3. Lead citrate: 0.3% lead citrate in double-distilled water.

2.5. Grinding and Polishing

1. SEM aluminum stubs, 13, 25, or 32 mm in diameter.
2. Silicon carbide paper for wet grinding: 180, 1,200, and 4,000 grit.
3. Polishing cloth and diamond paste with 3, 1, and 0.5 μm grain sizes.
4. EXAKT Grinding System.

2.6. Backscatter Calibration

1. MAC standard #6207 (containing aluminum and carbon).

2.7. Software

1. Photoshop Version 4 or higher.
2. SigmaScan Pro Version 5.
3. Excel.

3. Methods

3.1. Correlative LM and TEM Examination

As fixation, decalcification, dehydration, embedding, sectioning, and staining/contrasting are routine procedures in laboratories equipped for processing TEM specimens, the following list includes only the steps required when a resin block is ready for sectioning. Furthermore, details are described only for the steps, which are specific to the technique introduced here.

1. Cut a few survey sections from the resin block containing the entire specimen with a glass or diamond knife and stain them with toluidine blue. If necessary, adjust the plane of sectioning. Identify the area of interest and reduce the block size accordingly.
2. Using a histo diamond knife (or a glass knife), cut some sections of 1–2 μm in thickness and stain them with toluidine blue.

3. Using a histo diamond knife (or an ultra diamond knife, if the block is small enough), cut ultrathin sections of silver–gold interference color (80–100 nm) and collect them on coated grids. Try to place areas intended for TEM examination over gaps between grid bars.
4. Wearing latex gloves, contrast the sections with uranyl acetate and lead citrate.
5. Tranfer a grid to the TEM, turn on the appropriate high tension (we normally use 60 or 80 kV), and select the lowest possible magnification. Having located the area of interest identified in the LM, increase the magnification to the desired value (see Note 1). It may be advantageous to rotate the grid so as to align the grid bars approximately parallel to the *x*- or *y*-axis of the specimen stage.

Capture the micrographs

6. With the image acquisition software of the TEM camera, define the background correction (sometimes also referred to as shading) and determine the optimal exposure settings at the selected magnification. Then, identify one corner of the intended composite image and, starting from this point, collect contiguous rows and columns of micrographs by displacing the specimen stage only along either the *x*- or *y*-axis. Make sure that individual micrographs overlap by about 10%. Use constant exposure settings as determined at the start of capturing throughout the image acquisition.

Assemble the micrographs in Photoshop

7. Based on a rough estimate of the final size of the composite image, create a new document with a canvas size sufficient to accommodate the constituent micrographs (see Note 2).
8. Open a series of 10–12 constituent micrographs (see Note 3). Depending on their original resolution as well as the number of collected images, reduce the size of the constituent micrographs (e.g. to about 600–700 × 600–700 pixels) in order to avoid excess amounts of data for the composite image.
9. Select the first micrograph and adjust levels to range from about 5 (almost black) to 250 (almost white). Based on visual judgement, the brightness should be medium. With the Move tool, drag the micrograph to the appropriate corner of the canvas of the composite image, thereby creating a new layer. Label this layer for easy later identification.
10. Select the subsequent micrograph and drag it to a new layer of the composite image. Label this layer as well and, using the Zoom tool, zoom in to see only the two micrographs, which have to be superimposed. In the Layer menu, create a new

adjustment layer for level adjustment, make sure to activate Create Clipping Mask in the pop-up window (see Note 4), and adjust levels. If necessary, adjust also brightness and contrast in the same way, until the visual appearance of both micrographs correspond. Optionally, merge down adjustment layers to image layer (see Note 5).

11. Select the image layer and, using the glider in the Layers panel menu, reduce the opacity of the image to 50–60%, select the Move tool, and superimpose conspicuous structures in the overlapping portion of the two micrographs. Coarse movements are made with the mouse, and fine adjustments with the arrow keys. Restore the opacity of the image layer.
12. In the Layer menu, create a layer mask. With the Brush tool and black selected as foreground color, remove the margin of the micrograph, which is often slightly darker and irritating (see Note 6). If an irregular margin is thus created, the transition from one image to the next becomes less conspicuous.
13. Save the composite image, preserving all layers (see Note 7).
14. Repeat steps 10–13 for all remaining micrographs.
15. Crop the final image to the desired size and (optionally) flatten it to one level to reduce the amount of required storage space required.

3.2. Mineral Density Estimation

Carry out all procedures involving fixatives, solvents, and embedding media in a fume hood and wear latex gloves.

1. Fix specimens in half-strength Karnovsky's fixative or phosphate-buffered fixative (see Note 8); thereafter, rinse them thoroughly in wash buffer.
2. Depending on the intended plane of sectioning, create a flat specimen surface for embedding with a razor blade, diamond saw, or file.
3. Dehydrate the specimens in ascending grades of ethanol (one half-day, each, in ethanol concentrations from 50 to 95%, a full day in 100% ethanol).
4. Infiltrate with Technovit 7200 VLC (1 day in a 2 + 1 mixture of ethanol and Technovit, 2 days, each, in 1 + 1 and 1 + 2 mixtures, 1 week in 100% Technovit under vacuum of about 300 mbar). Avoid premature polymerization of the resin due to daylight irradiation by using brown glass vials and/or wrapping the vials with aluminum foil.
5. Embed specimens on the flat surface in Technovit 7200 VLC using molds of appropriate size. Avoid introducing air bubbles and place molds in vacuum for 5–10 min before polymerization.

6. Place molds in a cold water bath which is preferably cooled with running water, cover them with a thin Plexiglas plate, and polymerize the Technovit with white light for 2 h followed by blue light for 2 h (see Note 9).
7. Remove blocks from embedding molds (see Note 10) and glue them onto SEM aluminum stubs, taking care that the flat specimen surface is as parallel as possible to the stub surface.
8. Expose the area of interest of the specimen by wet grinding the block surface with silicon carbide papers of decreasing grit size. Visual controls with a stereomicroscope may be necessary to determine whether enough resin has been removed. When this is achieved, the block is polished with a polishing cloth and diamond pastes of grain sizes 3, 1, and finally 0.5 μm (see Note 11).
9. Let the blocks dry at room temperature for several days. Then, coat the polished surfaces with a 10–15-nm-thick layer of carbon.
10. Place a block together with aluminum and carbon standard samples in the SEM chamber (see Note 12). Select BSE mode and turn the high tension (20 kV) on. Let the filament stabilize for at least 15–30 min (see Note 13).
11. Select a working distance of 20–25 mm and leave it constant. Focus the standards and specimen only using the *Z*-drive of the microscope stage (see Note 14). Adjust gain and contrast of the signal so as to obtain a gray level of about 200–210 for aluminum and of about 5–10 for carbon (see Note 15).
12. Leaving the settings for gain and contrast as well as the working distance fixed, capture the necessary micrographs of the specimen (Fig. 1a–e) and then one micrograph, each, of the aluminum and carbon standard.
13. Open micrographs of the specimen in SigmaScan Pro. If deemed necessary, calibrate distances to change the scale from pixels to mm or μm. Determine gray-level values either as averages within an area of interest or at individual dots along a line. In the same way, record mean gray-level values from the micrographs of the aluminum and carbon standard. For this purpose, select Trace Measurements Mode and the following parameters from the Measurements Settings: Measurements Average Intensity or Line Intensity, the Column of the worksheet, where data are to be collected, an Overlay for the drawing, and Continuous (Streaming) Measurements for Trace mode (if Line Intensity has been selected, the Line Width in pixels is also defined here). Then, outline an area or draw a line using the left mouse key, and terminate drawing with a right mouse click. This automatically creates a worksheet containing the data in the selected column. Save this worksheet as an Excel file.
14. Make the necessary assumptions regarding the approximate chemical composition of the compound hard tissue (see Note 16)

and calculate the respective mean atomic numbers (MANs) and mean backscatter coefficients (MBSCs; see Note 17). We routinely use Müller's and Lloyd's formula for MAN and Castaing's formula for MBSC (18) (indicated as entered in an Excel formula):

$$\text{MAN} = \text{SUM}\ (c_i \times Z_i)\ \text{ for } i = 1 \text{ to } n \text{ constituent elements,}$$

where c_i = weight fraction of element i with atomic number Z_i

$$\text{MBSC} = \text{SUM}\ \left(c_i \times \text{BSC}_i\right)\ \text{ for } i = 1 \text{ to } n \text{ constituent elements,}$$

where c_i = weight fraction of element i with backscatter coefficient

$$\text{BSC}_i = 2^{(-9/\text{SQRT}(Z_i))}$$

Table 1 lists MANs and MBSCs useful for mineral density estimation.

15. Open the worksheet containing the raw gray-level data in Excel and derive the MBSC of the compound tissue ($\text{MBSC}_{\text{Tissue}}$). Then, calculate the mineral density estimates (MDEs; in weight%; see Note 18) as follows (indicated as entered in an Excel formula):

$$\text{MBSC}_{\text{Tissue}} = \begin{pmatrix} \text{BSC}_{\text{Al}} \times \left(I_{\text{Tissue}} - I_{\text{C}}\right) + \\ \text{BSC}_{\text{C}} \times \left(I_{\text{Al}} - I_{\text{Tissue}}\right) \end{pmatrix} / \left(I_{\text{Al}} - I_{\text{C}}\right)$$

where BSC_{Al} and BSC_{C} = BSCs of aluminum and carbon (Table 1).

I_{Tissue} = gray-level value obtained from compound hard tissue (0–255).

I_{Al} and I_{C} = gray-level values obtained from the aluminum and carbon standard.

If water is disregarded,

$$\text{MDE} = \left(\text{MBSC}_{\text{Tissue}} - \text{MBSC}_{\text{Matrix}}^{\text{a}}\right) / \left(\text{MBSC}_{\text{Mineral}} - \text{MBSC}_{\text{Matrix}}\right),$$

where $\text{MBSC}_{\text{Mineral}}$ and $\text{MBSC}_{\text{Matrix}}$ = MBSCs of mineral phase and organic matrix (Table 1).

If water is assumed to be present at a proportion of, e.g., 1 part matrix: 4 parts water,

$$\text{MDE} = \left(\text{MBSC}_{\text{Tissue}} - \text{MBSC}_{\text{Matrix}}\right) / 5 - 4 \times \text{MBSC}_{\text{Water}} / 5) / \left(\text{MBSC}_{\text{Mineral}} - \text{MBSC}_{\text{Matrix}} / 5 - 4 \times \text{MBSC}_{\text{Water}} / 5\right),$$

where $\text{MBSC}_{\text{Water}}$ = MBSC of water (Table 1).

Table 1
Mean atomic numbers (MANs) and mean backscatter coefficients (MBSCs) useful for mineral density estimation of hard tissues

Compound	MAN	MBSC
Tooth mineral $Ca_9[(PO_4)_{4.5}(CO_3)_{1.5}]OH_{1.5}$ (20)	13.871	0.179
Organic matrix (13)	6.4	0.085
Water	7.217	0.098
Aluminum	13.0	0.177
Carbon	6.0	0.078

4. Notes

1. In order to recognize ultrastructural details of individual cells, a minimum magnification of about 1,900–2,000× is required. Depending on the size of the area of interest, this may necessitate more than 100 micrographs.
2. The estimate of canvas size is not critical, as it can be adjusted at any time during assembling.
3. The advantage of opening several micrographs at a time is twofold. First, some of the subsequent steps can be carried out using batch processing and second, the original appearance of the images can readily be restored, if some of the adjustments made turn out to have been inappropriate.
4. If a clipping mask is not created, all adjustments made in an adjustment layer affect all images deeper down in the layer list.
5. Merging down adjustment layers reduces the length of the layer list which can become extensive and rather obscured, if large numbers of micrographs are assembled. The advantage of keeping the adjustment layers is that individual constituent images can be further modified (or restored to the original appearance) at any time.
6. When white is selected as foreground color, all changes made with the Brush tool can be reversed.
7. For easy identification, we use the Photoshop format (.psd) for the compound image including all layers and the TIFF format (.tif) for the flattened image.

8. The choice of fixatives is not critical. Depending on the size of the specimens, the duration of fixation can vary from 2 h to several days at room temperature or 4°C.
9. A convenient way of polymerization is provided by the EXAKT light polymerization unit with integrated cooling system.
10. Polymerization of Technovit 7200 VLC is impaired by oxygen from the air. Therefore, the top surface of the polymerized blocks is often rubberlike. When the blocks are stored at room temperature for 1–2 days and unpolymerized monomer is allowed to evaporate, the layer of soft resin can easily be scraped or ground away to produce a hard, rough surface for gluing.
11. In order to keep the block surface parallel to the stub, we grind blocks using the EXAKT grinding system. For this purpose, stubs are fixed in a custom-made holder which can be placed in the specimen holder of the machine like a microscope slide. Polishing is done by hand.
12. Instead of using a commercial product, standards can easily be prepared using small pieces of pure aluminum wire and graphite, which are embedded in Technovit 7200 VLC and processed exactly as the specimen blocks.
13. According to our experience, stabilization of the filament is one of the most critical factors for obtaining reproducible gray levels of the BSE signal. This is the reason that we use embedded rather than hydrated specimens in a low vacuum environment (environmental SEM conditions), in which specimens dry out quickly, making the results of the quantitative analysis unpredictable.
14. The working distance is also a very important factor because even small variations significantly affect the BSE signal intensity.
15. Modern digital microscopes usually offer the possibility to select an intensity histogram.
16. While it is clear that estimation of biological mineral densities has to take into account at least a mineral phase and the organic matrix, the question is whether to also consider water (which is even denser than organic matrix). When dehydrated and embedded material is analyzed, disregarding water may be reasonable. If it is to be considered, a further question is whether it is associated mainly with the organic matrix and/or in a certain fixed proportion, for instance the proportion found in normal, fully mineralized tissues. As shown in Fig. 3a, this assumption markedly affects the resulting estimates of mineral density, particularly when the contents of organic matrix are relatively high.
17. As revealed by Fig. 3a, using MANs instead of MBSCs has little impact on the resulting MDEs. Moreover, there is no

consensus as to the correct way of calculating MANs and MBSCs, and values obtained with the various methods differ considerably (18).

18. We do not calculate volume fractions because this would require still another assumption as to the specific weight of the tissue components.

Acknowledgments

The authors thank Jacqueline Hofmann-Lobsiger and Steven Reese for the skilful preparation of the specimens used for the illustrations of this work.

References

1. Pease DC (1964) Histological techniques for electron microscopy, 2nd edn. *Academic Press, New York.*
2. Hayat MA (1970-1978) Principles and techniques of electron microscopy: biological applications, vol 1–9. *Van Nostrand Reinhold Company, New York.*
3. Glauert AM (ed) (1972-1992) Practical methods in electron microscopy, vol 1–14. *North Holland, Elsevier, Amsterdam.*
4. Koehler JK (ed) (1973) Advanced techniques in biological electron microscopy. *Springer-Verlag, Berlin, Heidelberg.*
5. Hayat MA (2000) Principles and techniques of electron microscopy: biological applications. *Cambridge University Press, Cambridge (UK).*
6. Hayat MA (ed) (1987) Correlative microscopy in biology: instrumentation and methods. *Academic Press, Orlando (FL).*
7. Sutton-Smith P, Beard H, Fazzalari N (2008) Quantitative backscattered electron imaging of bone in proximal femur fragility fracture and medical illness. *J Microsc* 229:60–66.
8. Roschger P et al (1995) A new scanning electron microscopy approach to the quantification of bone mineral distribution: Backscattered electron image grey-levels correlated to calcium K α-line intensities. *Scanning Microsc* 9:75–88.
9. Bloebaum RD et al (1997) Determining mineral content variations in bone using backscattered electron imaging. *Bone* 20:485–490.
10. Skedros JG et al (1993) The meaning of graylevels in backscattered electron images of bone. *J Biomed Mater Res* 27:47–56.
11. Skedros JG et al (1993) Influence of mineral content and composition on graylevels in backscattered electron images of bone. *J Biomed Mater Res* 27:57–64.
12. Vajda EG, Skedros JG, Bloebaum RD (1995) Consistency in calibrated backscattered electron images of calcified tissues and minerals analyzed in multiple imaging sessions. *Scanning Microsc* 9:741–755.
13. Boyde A et al (1992) Applications of mineral quantitation of bone by histogram analysis of backscattered electron images. In: Slavkin H, Price P (eds) Chemistry and biology of mineralized tissues. *Elsevier Science Publishers B.V., Amsterdam*, p 47.
14. Angker L et al (2004) Correlating the mechanical properties to the mineral content of carious dentine – a comparative study using an ultra-micro indentation system (UMIS) and SEM-BSE signals. *Arch Oral Biol* 49:369–378.
15. Angker L et al (2004) Quantitative analysis of the mineral content of sound and carious primary dentine using BSE imaging. *Arch Oral Biol* 49:99–107.
16. Mahoney EK et al (2004) Mechanical properties and microstructure of hypomineralised enamel of permanent teeth. *Biomaterials* 25:5091–5100.
17. Hayat MA (1981) Fixation for electron microscopy. *Academic Press, New York.*
18. Howell PGT, Davy KMW, Boyde A (1998) Mean atomic number and backscattered electron coefficient calculations for some materials with low mean atomic number. *Scanning* 20:35-40.
19. Harada H, Ohshima H (2004) New perspectives on tooth development and the dental stem cell niche. *Arch Histol Cytol* 67:1–11.
20. Driessens FCM, Verbeek RMH (1990) *Biominerals.* CRC Press, Boca Raton (FL), p 131.

Chapter 10

Deoxyoligonucleotide Microarrays for Gene Expression Profiling in Murine Tooth Germs

Harald Osmundsen, Anne-Marthe Jevnaker, and Maria A. Landin

Abstract

The use of deoxyoligonucleotide microarrays facilitates rapid expression profiling of gene expression using samples of about 1 μg of total RNA. Here are described practical aspects of the procedures involved, including essential reagents. Analysis of results is discussed from a practical, experimental, point of view together with software required to carry out the required statistical analysis to isolate populations of differentially expressed genes.

Key words: mRNA, Transcriptome, Hybridization, Fluorescence, Cy3, Cy5, MA-plot, Clustering, .gal-file

1. Introduction

1.1. The Microarray Technology

Experimental measurements of gene expression were, prior to the advent of high-throughput technologies, limited to a few genes at a time due to the amount practical work involved using for example Northern blotting.

We have carried out gene expression profiling at various developmental stages of the murine first molar tooth germ or of a section of an incisor (1–3). Microarray analysis of gene expression can be used to isolate genes, which are differentially expressed at different developmental stages. This analysis requires the use of appropriate techniques for detection of hybridization signals as well as statistical software to analyze data. In the following sections, practical techniques involved are described in a manner which hopefully should enable the reader to undertake similar experiments. The procedure described is likely more suitable for a research laboratory, rather than a laboratory carrying out a large volume of routine analysis.

Chrissa Kioussi (ed.), *Odontogenesis: Methods and Protocols*, Methods in Molecular Biology, vol. 887,
DOI 10.1007/978-1-61779-860-3_10, © Springer Science+Business Media, LLC 2012

About 15 years ago, the microarray technology emerged (4), fundamentally altering the analysis of gene expression. A technique enabling simultaneous assays of expression of hundreds, and later thousands of genes, was at hand (5). Currently, microarray slides with 30,000 probes covering most known genes of the human or murine genome are commercially available (6–8). Slides with probes representing most of the genes of the rat or yeast genome are also commercially available. Slides for other species may be available from independent manufacturers, e.g. academic core facilities equipped with slide printers.

The original principle of microarray design, the covalent attachment of probes on to the surface of the glass slide, remains in use today, see review (7). It is, however, appropriate to point out that alternative techniques are now also available, e.g. Affymetrix (Santa Clara, CA, USA), Illumina (San Diego, CAL, USA), and Nanostring Technologies (Seattle, WA, USA). Although the more recent technologies, holding great promise and represent significant technological advances, the original glass slice technique nevertheless remains useful. Because it is an open technology, not licensed to any one manufacturer, it is invariably available at a lower cost. Typically, a microarray glass slide, ready for use, may be purchased for about US $100 each. We have found that microarrays based on classical printed microarrays provide a cost-effective approach to microarray analysis. Such slides may be purchased from for example Agilent (Santa Clara, CA, USA) or Phalanx Biotech (Palo Alto, CA, USA). Several academic core-facilities also print microarray slides of competitive quality.

The use of classical microarray slides also requires the use of an appropriate detection technology. Fortunately, several of these are commercially available: e.g. 3DNA array expression detection (Genisphere Inc, Hatfield, PA, USA), HiLight Array detection System (Qiagen, Hilden, Germany), CyScribe labelling kit (GE Healthcare, Bucks, UK), ULSlabelling kits (Kreatech Diagnostics, Amsterdam, Holland). In the following section, the use of the Genisphere's 3DNA™ array expression detection kits is described because these kits, in our hands, have been found to yield good results (Fig. 1). This does not, however, preclude that alternative detection kits would have performed equally well.

Microarray analysis also requires an appropriate microarray scanner, which is available from various manufacturers. Classical glass microarray slides cannot be used with scanners from for example Affymetrix (Santa Clara, CA, USA) or Illumina (San Diego, CA, USA). However, scanners from a range of alternative manufacturers may be used (e.g. Agilent Technologies, MA, USA), Axon (Sunnyvale, CA, USA), CapitalBio Corporation (Beijing, China), GE Healthcare, INNOPSYS (Carbonne France), Packard BioScience (Perkin-Elmer, Waltham, MA, USA), Roche Applied Science (Indianapolis, IN, USA).

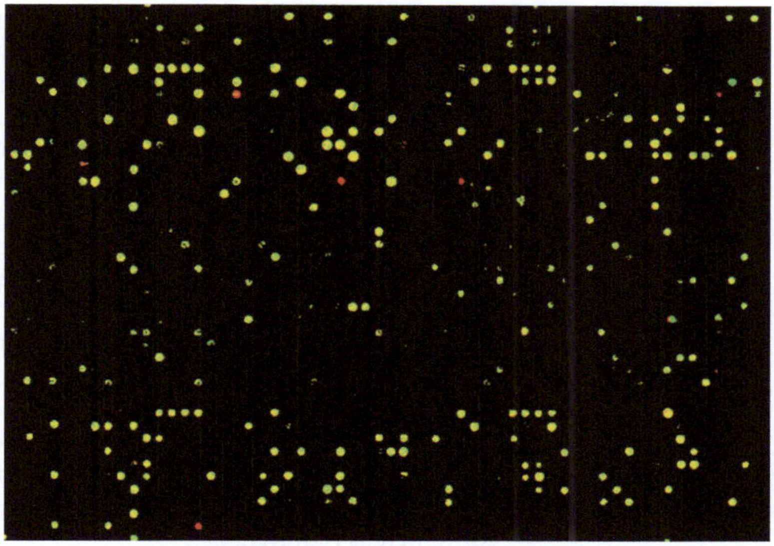

Fig. 1. Typical composite image of a microarray where hybridization was detected using the Genisphere 900 detection kit and a 30 K murine microarray from Phalanx Biotech (CA, USA). The image shows a fraction of the microarray image and was obtained with RNA isolated from first molar tooth germs removed at E15.5 (Cy3) and P1 (Cy5). The composite image is obtained by superimposing the Cy3 and Cy5 images. Most of the spots represent genes which are expressed in both samples, hence the *yellow colour*, while a few spots are predominantly *green or red*, indicating much higher expression of those genes at E15.5 or P1, respectively.

Microarray data requires statistical analysis to isolate populations of differentially expressed genes. This is invariably the more time-consuming component of microarray work, and availability of user-friendly software is therefore most desirable, including academic BRB ArrayTools, ViDaExpert (http://www.bigre.ulb.ac.be/Users/jvanheld/web_course_microarrays/) and commercial software (Gene Spring, Agilent Technologies), Partek Genomics Suite (Partek Inc., St. Louis, MISS, USA), Spotfire DecisionSite for Microarray Analysis (Tibco Spotfire, Somerville, MA, USA). In general, commercial software is often significantly more user-friendly, although not free of charge. The cost of a licence should, however, be balanced against time spent learning to use less user-friendly software.

1.2. Microarray Analysis of Gene Expression

During murine tooth development, substantial developmental changes occur even within a time-span of 24 h. From embryonic day (E) E11.5 up to post-natal day (P) P5, i.e. in the course of about 16 days the tooth germ develops from oral epithelium into a phenotypic molar tooth exhibiting extensive mineralization. The developing murine tooth germ is therefore an excellent model for studying cell differentiation. We have shown that almost 1,400 genes are differentially expressed between E15.5 and P2, suggesting that a substantial number of genes are involved during this process (1–3). A large fraction of these genes are have not been previously mapped, and many of the genes have unknown functions.

It is likely that ongoing cellular activities are a reflection of the current composition of the mRNA transcriptome. By mapping changes in the global mRNA transcriptome during development, the resulting data can therefore be used to identify predominant cellular activities at different developmental stages. Such mapping of the global mRNA transcriptome requires the use of high-throughput technology, e.g. microarrays. Typically, dual labelled samples of mRNA are used facilitating analysis of two separate samples of RNA on a single slide. For subsequent statistical analysis, however, the resulting data may be treated as single channel data, provided the software used facilitates single channel statistical analysis. This approach provides substantially more flexibility as regards analysis of data because sets of data from different slides can be combined as required.

2. Materials

2.1. Water

It is imperative that all solutions are prepared using RNAse free water. This may be purchased (e. g. Accugene Molecular Biology Water, Lonza Inc., Rockland, Me, USA) or prepared using diethylpyrocarbonate to ensure irreversible inactivation of ribonuclease (9). This is absolutely essential for a successful investigation. Prepare all solutions at room temperature, and store at −70°C, unless otherwise indicated.

2.2. RNA Extraction

For successful microarray analysis isolated mRNA should not be significantly degraded. Additionally, the procedure for isolation of total RNA must be selected with the labelling procedure in mind. Indirect labelling procedures usually involve the use of enzymes and are therefore sensitive to remaining organic molecules if phenol-based extractions are used. Solvent-extraction based kits cannot, therefore, be used together with an indirect labelling kit. Therefore, a column-based procedure (Qiagen RNeasy mini kit, Qiagen, Hilden, Germany) should be used together with an indirect labelling kit (Genisphere 900). RNA concentrations should be calculated (OD260/280 1.8–2.0. and OD260/230 >1.7). Further characterization using an Agilent Bioanalyzer (Agilent, Palo Alto, CA, USA) is preferable. Solutions of RNA should yield RIN-values above 8.5 for use in microarray analysis. Solutions of total RNA exhibiting these characteristics should therefore be of both high purity and exhibit extensive preservation of structural integrity and therefore be well suited for microarray work.

2.3. Isolation of RNA from Tooth Germs

Removal of tooth germs from mouse embryos is best carried out using a suitable stereomicroscope, especially at pre-natal stages. Embryos are instantly killed by decapitation and the head is submerged in RNAlater (Ambion Inc., TX, USA) diluted 1:3 with PBS at RT to preserve RNA integrity. All subsequent dissection is carried out with the head being submerged in this medium. Once removed the tooth germs are immediately transferred to undiluted RNAlater (at RT), shaken gently, and left at RT for at least 1 h prior to storage at −20°C. At the most five tooth germs should be added to 2 ml of RNAlater.

Using the RNeasy Mini kit (Qiagen, Hilden, Germany) RNA is isolated using batches of five pre-natal tooth germs, while RNA may be isolated from single post-natal tooth germs. Tooth germs are transferred to 50 μl of homogenization medium (supplied with the RNeasy Mini kit), and homogenization of tooth germs is carried out using a small Ultraturrax knife (5 mm diameter) for about 10 s. More convenient, and reproducible, homogenization is carried out with a Precellys homogenizer (Bertin Technologies, Orléans, France), using 100 mg of zirconium oxide beads added to tooth germs contained in 30–50 μl of homogenization medium. Tooth germs are fully homogenized following treatment for at no more than two cycles, each of 10 s duration, in the Precellys homogenizer. Subsequent purification of total RNA is carried out as described in the RNeasy Mini manual.

2.4. Microarrays

Microarrays were initially printed using cDNA probes. Today, however, deoxyoligonucleotides is the molecule of choice for probes. Deoxyoligonucleotides 50–70 bases long are invariably used, the length of the probe may vary somewhat depending on the probe set used. It is advisable to ascertain that the purchased microarrays have been printed using a high quality deoxyoligomer probe set (10), e.g. using probe sets from Operon, MEEBO, RefSeq release 38, Ensemble release 56. The microarray should also contain spots with probes for an appropriate selection of external controls (see below).

2.5. Design of Microarray Studies

A microarray study should have a clear goal and well-defined, precise, hypothesis for testing. These points should be established prior to initiating experimental work to avoid carrying out unnecessary, expensive, experiments.

To avoid systematic bias, the design of a microarray experiment should take into account all sources of variation. RNA samples ought to have the same quality, and all samples of RNA used should be processed in the same manner. It is also good practice to use a strain of experimental animals exhibiting minimal genetic diversity to minimize biological variation, although and out bred strain of mice, e.g. CD1, should suffice. The isolation of statistically significant, differentially expressed, genes may otherwise prove difficult.

A sufficient number of replicates are invariably required to demonstrate significantly differentially expressed genes. The number of replicates required depends on extents of biological and technical variation. In microarray work at least triplicate measurements are invariably required. This may require a number of separate biological samples, and a correspondingly high number of microarrays. The cost may therefore escalate to a prohibitive level. As a cost-cutting compromise, although statistically undesirable, samples of RNA isolated separately from three to five different individuals may be pooled. The pooled RNA is analyzed using at least three separate microarrays.

2.6. Microarray Hybridization

Microarray hybridization is carried out on a glass slide using 50–60 μl of hybridization medium. Allowing a regular coverslip to float on the top of the hybridization medium could cause very uneven hybridization of mRNA to the probes on the slide. This problem is eliminated by using lifter slips (Electron Microscopy Sciences, Hatfield, PA, USA), i.e. coverslips having a raised edge. A volume (e.g. 20 μl) of hybridization medium is positioned onto the edge of the pre-positioned lifter slip (using a pipette with a long tip), capillary forces subsequently ensuring that the open space between lifter slip and array surface is filled with hybridization medium. Capillary attraction will ensure that the lifter slip remains in position during subsequent hybridization.

Hybridization of microarray slides can be carried out in small, individual, hybridization chambers submerged in a thermostatted water bath with gentle shaking. This is, however, far from ideal. Uneven hybridization (i.e. patches on the microarray which has not been adequately hybridized therefore exhibiting weak, or absent, hybridization) will invariably occur. This problem can be successfully eliminated by employing a hybridization station (e.g. SlideBooster 400 Hybridization Station, Advalytix, Munich, Germany). The SlideBooster will by means of surface acoustic waves ensure effective agitation of the hybridization medium during hybridization and therefore facilitating even hybridization across the entire array. The SlideBooster also include a slide-washer, which we have found most convenient for post-hybridization washing of slides.

2.6.1. Selection of Kit for Detection of Microarray Hybridization Signals

Microarray hybridization signals may be detected using either direct or indirect labelling. Direct labelling procedures (e.g. CyScribe labelling kit) are usually less time-consuming than indirect labelling procedures (e.g. Genisphere 900 3DNA array expression detection kit), the latter requiring one additional hybridization. One advantage of indirect labelling is the absence of differences in labelling efficiencies between Cy3 and Cy5. Alexa-dyes can also be used for detection, although these dyes, like Cy3 and Cy5, are unstable and very sensitive to photo-bleaching.

Microarray hybridization requires cDNA synthesis using mRNAs contained in the purified total RNA as templates. When using the Genisphere 900 kit for detection of hybridization signal the oligo-dTprimer contained in this kit must be used for cDNA synthesis. This primer also contain a 3′-terminal 3DNA capture sequence which is essential for subsequent detection, as the resulting cDNAs will have an attached 3′-capture sequence to which the chromophore (Cy3 or Cy5)-containing DNA will bind. This procedure therefore includes two stages of hybridization: Initial hybridization of cDNAs to probes on the glass slide, and a subsequent hybridization to facilitate binding of 3DNA-chromophore to capture-sequence on cDNAs hybridized to probes attached to glass surface of the array. In our hands, the Genisphere Array 900 kit were invariably preferable as successful hybridization and detection routinely is carried out using 0.5–1 μg total RNA per sample per array. Working with tooth germs, this is desirable, as small amounts of RNA are often available, especially at early developmental stages. With tissues facilitating isolation of higher amounts of RNA Array 50 or Array 350 kits from Genisphere can be used.

2.6.2. Scanning of Slides

When using printed microarrays it is essential to ensure that the microarray scanner settings are such that most spots give fluorescence signals within a linear response range, i.e. as few spots as possible do not yield saturating signals. An appropriate laser power setting is essential in this respect. A laser power setting of 90% (of maximal power) is often used, but this may have to be decreased towards 80% if many spots exhibiting saturating fluorescence signals are apparent.

Linearity of the fluorescence signal response must also be ascertained. Printed slides are likely to have spots containing probes for external controls (11). Arabidopsis thaliana probes can be used in this respect, requiring the purchase of Arabidopsis thaliana mRNA spikes (Stratagene, Agilent Technologies Inc., Santa Clara, CA, USA). These, or similar, external controls provides an excellent tool for checking the specificity- and sensitivity of hybridization as well as quality of the fluorescence signals in the Cy3 and Cy5 channels. The OneArray slides (Phalanx Biotech Group, Palo Alto, CA, USA) contain a range of spots with external control probes allowing the user to quickly optimize scanner settings and to examine specificity and sensitivity of hybridization. Slides without appropriate external control probes are best avoided.

2.6.3. Quantitation of Microarray Fluorescence Signals

Following successful hybridization and scanning the resulting hybridization signals are quantitated. To this end, most suppliers of scanners will supply software suitable for quantitating scanned images of slides (usually as high resolution 16-bit tif-files). We have found ScanArrayExpress (v. 3) software (Perkin Elmer Life Sciences) most useful in this respect. Instrument-independent software is, however, also available, e.g. Array-Pro Analyser (Media Cybernetics Inc., Bethesda, MD, USA), Imagene (BioDiscovery Inc, El Segundo, CA, USA), GenePix Pro 7 (Molecular Devices Inc., Sunnyvale, CA, USA).

One further bit of software is required for successful quantization of slide data—an appropriate .gal-file, which is usually supplied by the microarray manufacturer. This file defines a grid of spots, which should mirror the microarray layout. It also contains the coordinates together with a unique identifier for each spot on the array. The .gal-file format is, however, not universal. In our experience, some of the gal-files routinely supplied by the slide manufacturer will not work with every type of scanner. In practice, this is reflected by the inability of the gal-file to position a grid correctly on the scanned image. Hence, it is advisable to ask the manufacturer of the slides if the gal-file supplied has been tested with the particular make of scanner, which is available. An incompatible .gal-file must be replaced—the slide manufacturer should be able to solve this problem.

It should be remembered that most software used to quantitate fluorescence signal intensities also should be able to carry out normalization of quantitated data, usually using for example LOWESS normalization (12). Normalization attempts to remove variability between sets of data, which is unrelated to biological variation. Normalization of data within each array is important to control for systematic biases in dye coupling, hybridization efficiencies, technical biases in the probes and the print tip used to manufacture the array. Sometimes, one channel is found to yield signals of much higher intensity that the other channel because the labelling efficiency was much better in one channel. This will be corrected using normalization. The resulting data file, should, however, record both normalized and non-normalized (raw) fluorescence intensities.

The raw signal intensity for each experimental set of probes should show a uniform distribution of signals, i.e. plotting of Cy5 (red channel) intensities against Cy3 (green channel) intensities should yield a linear regression curve having a slope the slope about 1. The quality of the quantitated fluorescence data can be further ascertained by examining the resulting MA-plot (12). The MA-plot is constructed by plotting the red/green intensity ratios ("*M*") against the average intensities ("*A*") for each spot on the microarray. The "*M*" is defined as $\log_2$ (Cy5 signal intensity/Cy3 signal intensity). "*A*", is a measure of signal intensity being defined as ($\log_2$Cy5 signal intensity + $\log_2$Cy3 signal intensity)/2. The *M* and *A* values should be calculated for all spots in microarray. It is advisable to examine MA-plots prior to proceeding with further analysis of data. Correctly quantitated fluorescence intensities should gen-

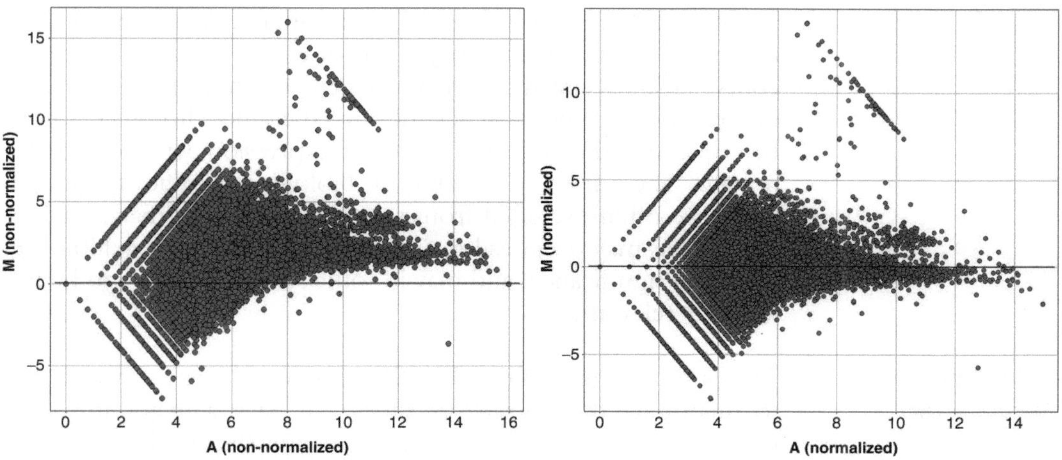

Fig. 2. MA-plots using non-normalized and normalized microarray data. Murine 30 K microarrays (Phalanx Biotech) were hybridized with RNA isolated from first molar tooth germs of mice. Labelling was carried out using the Genisphere 900 labelling kit. The MA-plot obtained when using non-normalized data exhibit a marked "banana" shape which is appreciably less marked when with normalized fluorescence intensities, suggesting uneven labelling in the two channels (Cy3 and Cy5) not due to biological differences. The use of normalized data is therefore recommended. It is nevertheless advisable to ensure this to be the case prior to proceeding to further statistical analysis of the results. The series of points gathered as a straight line above the middle of the plot are due to control spots present in the array.

erate a linear MA-plot with a gradient close to zero. Non-linear MA-plots suggests that normalized intensities should be used. In this case normalization (e.g. Lowess normalization) may linearize the MA-plot, also suggesting that Lowess-normalized data should be used for statistical analysis of data. Often both non-normalized and normalized data may yield closely similar MA-plots, suggesting microarray data of good quality. MA-plots derived from analysis of gene expression in the first molar tooth germ is presented in Fig. 2, and illustrate how the use of LOWESS normalized data corrected a systematic difference in fluorescence intensities probably caused by differences in labelling efficiencies in the Cy3 and Cy5 reactions. The curved, "banana-shaped" MA-plot is substantially linearized when LOWESS normalized data are plotted.

Appropriate pre-processing of microarray data is critical for identifying differentially expressed genes. A sample of total RNA will invariably contain mRNAs at a wide range of concentration. Low concentration mRNAs will yield fluorescence intensity signals lower than a predefined threshold (e.g. an intensity value of 400 or less). These will therefore invariably be classified as not detected because low intensity genes are often found to exhibit large variation between technical replicates. Signal intensities below the preset level are therefore often not included in any subsequent analysis as these likely to be highly unreliable. Conversely, some mRNA is likely to be present at very high levels and may therefore yield saturating fluorescence signals.

Because of their relatively high variance between technical replicates most low intensity genes are likely eliminated in the course ensuing statistical analysis when searching for significantly differentially expressed genes. In our experience, however, a number of genes showing low levels of expression will nevertheless remain. These may be of doubtful biological significance and are therefore often excluded from subsequent bioinformatic analysis. Exceptions may occur as low intensity genes may also exhibit high statistical significance, and may in some situations prove to be biologically relevant.

2.6.4. Statistical Analysis of Microarray Data

The output format of the data-file generated by the software used to quantitate the microarray slides can have a major impact as regards your choice of software used for statistical analysis. By default most programs will output your quantitated microarray data as .gpr-files. Unfortunately, however, .gpr-files generated by any two programs are likely to be sufficiently different to be incompatible. When using a slide-dedicated technology, e.g. Affymetrix or Illumina, this is unlikely to be a problem, as most software has import filters for their .gpr-files. A file from an independent slide-manufacturer, however, is unlikely to have a working import-filter in your software. Well-written software, however, allows data input using copy and paste from for example a .csv- or a .gpr-file.

The software chosen for statistical analysis of your microarray data must therefore have some alternative means of import of data. The data file is opened using for example Excel or Notepad. The ease by which this is carried out, however, may vary. In our hands Spotfire DecisionSite for MicroarrayAnalysis (TIBCO Spotfire, Somerville, MA, USA), or the more recent Integromics Biomarker Discovery (TIBCO Spotfire), are most convenient in this respect. Alternative software, however, will also do so, e.g. GeneSpring (Agilent, Lexington, MA, USA), Qlucore (Qlucore AB, Lund, Sweden). Excellent academic, licence free, software is also readily available via Internet, e.g. the Bioconductor package (http://www.bioconductor.org/packages/release/Software.html) or the TM4 microarray software suite (http://www.tm4.org/). Invariably data input using commercial software is markedly simpler and will invariably offset the cost of purchasing a software licence.

2.7. Statistical Methods

A range of statistical data analysis methods has been developed for gene expression profiling. Frequently used methods can be classified as differential analysis, e.g. *t*-test, ANOVA (13), supervised classification methods (linear discriminant analysis), dimensional reduction (principal component analysis) and unsupervised cluster analysis (hierarchical clustering, *K*-mean clustering, self-organizing maps). Differential analysis entails statistical analysis in order to discover differentially expressed genes by estimating the significance of differences between experimental mean values. Classification and cluster analysis are analytical approaches that attempt to decrease the dimensionality of microarray data by trying to sort records into similar classes. A supervised method makes use of predefined classes. Cluster analysis, in contrast, is referred to as a class discovery method, since no classes are predefined. This method groups genes into classes based on their similarity in response to experimental treatment or time of development. The resulting clusters have no biological meaning, but are most useful as a means of showing trends in changes of gene expression. The resulting clusters may, however, be most useful for successive bio-informatic analysis of data. Clusters of genes exhibiting different responses to treatment, or a different developmental time-course of expression, may yield worthwhile results when subjected to bio-informatic analysis. An extensive discussion of these topics is beyond the scope of this text. Several books, however, provide an introduction to the topics, e.g. (14). Most of these statistical techniques should also be available in software used for analysis of microarray data (see preceding section).

2.7.1. False Discovery Rates

Detection of differentially expressed genes requires statistical analysis to determine the significance of differences between population means, e.g. two-ways, or multiple, ANOVA (13). This is readily available in the various software used for analysis of microarray data, as are alternative statistical tests, e.g. Wilcoxon-test, Mann-Whittney test. The combined use of ANOVA together with FDR should ensure a robust selection of differentially expressed genes from most sets of microarray data.

Because analysis of microarray data involves the simultaneous testing of thousands of hypothesis there is a need to correct for false positive results among genes judged to be differentially expressed. Adjustments of *P*-value, based on controlling the probability of finding at least one false positive value, as is the case with for example ANOVA, will now be too conservative. A list of differentially expressed genes may therefore contain numerous false positives. The false discovery rate (FDR) of a data set is defined as the expected proportion of false positives among the differentially expressed genes (15, 16). In practice, a *P*-value is computed for each gene in the data set, and the level of the FDR is selected by selecting an appropriate level of the P-value required for a gene to be judges as statistically true positive. Usually, a value of $P \leq 0.05$ is used, but a lower value is often more appropriate if the data exhibit substantial variability.

2.7.2. Validation of Microarray Results

Validation of microarray results using for example real-time RT-PCR is essential. Only when changes in levels of gene expression observed using microarray analysis are shown to be similar to those obtained using real-time RT-PCR analysis can microarray results be considered reliable. The use of a standard cyber green

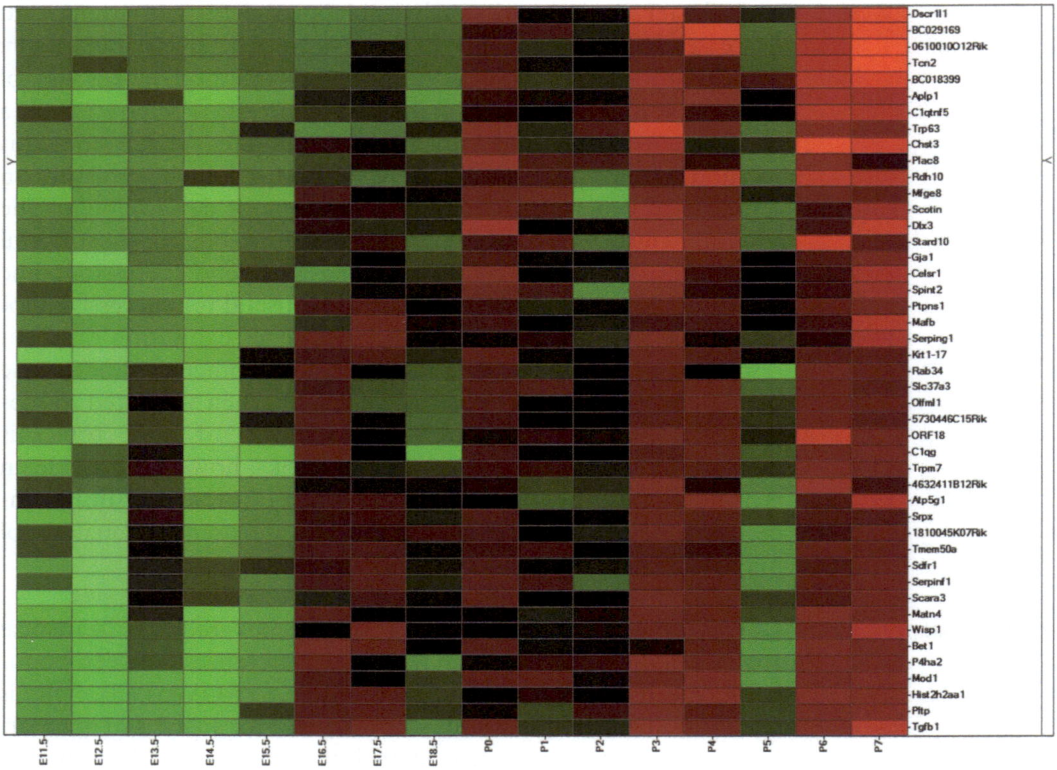

Fig. 3. Time-course of gene expression in first molar tooth germs isolated at 24 h intervals starting at E11.5, ending at P7. Results from 118 microarrays were assembled into a single data file for statistical analysis using Spotfire Decision Site Microarray Analysis software. Levels of expression at each developmental time-point are means derived from 3 to 5 microarrays. Log2 fluorescence intensities were subjected to FDR analysis (15), and multiple ANOVA to isolate differentially expressed genes. Profile search was carried out on differentially expressed genes to isolated genes with a time-course expression profile similar to that of Dlx3. This resulted to the isolation of 45 genes, which were subjected to hierarchical clustering yielding the heat diagram. Prior to hierarchical clustering log2 fluorescence intensities were *Z*-score and 0–1 normalized to minimize non-biological variation among results from the various microarrays.

assay will suffice, and results may be analyzed using a recent version of REST software (17). This is not to say that data obtained by the two methods should be identical in terms of extents of changes in gene expression. The changes observed should, however, exhibit identical directions of change, although some differences in magnitude of change are acceptable.

The final goal of microarray analysis is to present reliable clusters of differentially expressed genes which can be used in further work, e.g. for bioinformatics or for analysis of proteins. Results presented in Fig. 3 show a cluster of differentially expressed genes isolated from a time-course study of gene expression during development of the murine first molar tooth germ. There the data associated with these genes was used hierarchically cluster the genes, yielding the heat diagram shown in the Fig. This can often be a convenient method for visualizing time-course data.

3. Methods

In the following section, the various stages of a typical routine for cDNA synthesis and hybridization and labelling are outlined. Prior to starting practical work, it is good practice to ensure that all required reagents and equipment (as suggested by the manufacturers manual) are available.

1. Samples of RNA (about 1 μg of RNA/μl) of high purity and high degree intactness are essential. Dissolved RNA should be stabilized using a ribonuclease inhibitor (e.g. Superase.In, Applied Biosystems/Ambion, TX, USA) and stored in RNA storage solution (Applied Biosystems/Ambion) at −70°C. Concentrating solutions of RNA using a centrifugal evaporator should be avoided as this may lead to poor labelling efficiency. If concentration solutions of RNA is required, conventional precipitation using alcohol, followed by re-solubilization in a minimal volume of water (containing ribonuclease inhibitor) is recommended.
2. When designing experiments the use of biological replicates is preferable, and in classical microarray work triplicates. This, however, renders microarray experiments highly voluminous both in terms of data and cost. The problem may be circumvented by mixing purified RNA from biological triplicates (1:1:1, in terms of μg RNA) and running triplicate microarray analysis of the pooled RNA. The use of 1 μg of RNA pr sample per array is highly recommended. Lower amounts results in loss of detection signal intensity while a higher amount is unlike to significantly improve levels of detection.
3. Reverse transcriptase enzyme may be purchased from most sources (e.g. Superscript II, Gibco Cat. No. 18064-0, or corresponding product from Promega or Applied Biosystems). In our hands RevertAid M-MuLV Reverse transcriptase (EP0442, Fermentas GmbH, St. Leon-Rot, Germany) has performed well.
4. All operations should be carried out in a sterile-bench to avoid any contamination by ribonuclease. Likewise, pipettes used should be stored in a sterile bench and should be used exclusively for work with solutions of RNA of high purity.
5. It is often preferable to pre-wash microarray slides in 96% ethanol for 2–4 min to remove remaining traces of reagents used during printing of the slide. This will diminish background colouring and eliminate "comet-tail" spots. The slides must be thoroughly air-dried prior to use.
6. For cDNA synthesis and the use of 1 μg RNA per reaction mixture has been found suitable. This implies that 1 μg of

purified RNA should be used for each labelling reaction, for Cy3 and Cy5, respectively.

7. Hybridization must be performed using optimized temperatures and times for both hybridization and washing. These parameters depend on the labelling kit and the hybridization buffer used. Incubation temperatures that are too high will result in decreased signal intensities, while too low temperature will result in cross-hybridization, unspecific binding and apparently high signal intensities. The hybridization chamber must be properly sealed to maintain constant humidity throughout the hybridization. It is also important that the hybridization solution completely covers the array surface under the coverslip. The hybridization time depends on the labelling kit and the labelling procedure being used. The Genisphere 900 kit hybridization of labelled samples of RNA was carried out at 54°C for 18 h in a SlideBooster 400 Hybridization Station (Advalytix, Munich, Germany).
8. Transfer the slide quickly between wash steps, and centrifuge immediately after the last washing step to quickly dry the array. Do not expose the slide to air between washes for more than a few seconds.
9. Washing times and washing buffers are defined from the choice of labelling kit. It may often prove worthwhile to run a sample at temperatures 1–5°C higher or lower than recommended hybridization/washing temperatures to ensure that procedure is optimized with respect to the samples.
10. It is important to avoid direct exposure of the hybridized array containing the fluorescent dyes to light, as these dyes are readily photo-bleached. Hybridization and wash procedures should be performed in darkness or at least low-light conditions.
11. The array should be scanned within 30 min. after hybridization. Our microarrays were scanned in a Packard BioScience ScanArray Lite microarray scanner (PerkinElmer Life and Analytical Sciences Inc., MA, USA). The fluorescence signals were quantified using the ScanArray Express v. 3.0 TM program (PerkinElmer Life and Analytical Sciences, Inc.). The resulting fluorescence data, contained in a csv-file, was analyzed using the Spotfire v. 8.0TM DecisionSite for Microarray Analysis (TIBCO Spotfire, Somerville, MA, USA).
12. Statistical analysis is carried out using appropriate software to isolate differentially expressed genes.

The reader should consult the manufacturer manual for further details as these may differ with a recently purchased labelling kit.

4. Notes

1. The use of an RNA isolation kit not based on organic solvents is strongly recommended. Trace organic contaminants may interfere with subsequent cDNA synthesis or labelling.
2. The importance of using RNAase free solutions cannot be over-stressed, especially with purified fractions of RNA. RNase inhibitor should be included in all solutions; an RNase inhibitor not requiring added DTT is preferable.
3. Reserve pipettes exclusively for RNA work. Do not use electronic pipettes for pipetting solutions for labelling reactions as electronic pipettes may have inadequate precision.
4. All work entailing solutions of purified RNA is best carried out in laminar air-flow cabinets. This applies in particular to operations relating to labelling of cDNA for microarrays.
5. Always test your procedures using a single microarray slide expensive mistakes can be avoided in this manner.
6. Sealed microarray hybridization chambers, requiring incubation in a water-bath, are available at low cost. Their use may be appropriate when you are starting a microarray project and will enable you to verify that you procedures are working. These chambers invariably do not yield satisfactory hybridization, i.e. the efficiency of hybridization is likely to vary across the microarray slide. The use of a hybridization station will ensure homogenous hybridization across the slide, and therefore markedly improved the reliability of your data. The use of a hybridization stations is therefore strongly recommended.
7. Although commercial microarray slides usually come with an expiry date, you may find slides performing well for up 1 year past expiry provided they have been stored dry and in darkness. As slides age beyond expiry, however, you may find that hybridization signal intensities will decrease.
8. Replicates are essential in microarray work, use at least triplicates.
9. Rapid scanning after completion of slide processing is essential. Remember to scan Cy5 prior to Cy3.
10. Remember that statistical analysis of data microarray slides invariably will be the most time-consuming part of your microarray project. The application of user friendly software will therefore save huge amounts of time and effort. Unfortunately, the more user-friendly software is invariably commercial, but a software package you feel comfortable with is well worth this investment.

Acknowledgements

The expert technical assistance of Mrs. Toril Woldene, Mrs. Bente Gehrken, and Mr. Benedicto Geronimo are gratefully acknowledged.

References

1. Osmundsen, H., Landin, M. A., From, S. H., Kolltveit, K. M., and Risnes, S. (2007) Changes in gene-expression during development of the murine molar tooth germ, *Arch Oral Biol 52*, 803–813.
2. Sehic, A., Khuu, C., Risnes, S., and Osmundsen, H. (2009) Differential gene expression profiling of the molar tooth germ in peroxisome proliferator-activated receptor-alpha (PPAR-alpha) knockout mouse and in wild-type mouse: molar tooth phenotype of PPAR-alpha knockout mouse, *Eur J Oral Sci 117*, 93–104.
3. Sehic, A., Risnes, S., Khan, Q. E., Khuu, C., and Osmundsen, H. (2010) Gene expression and dental enamel structure in developing mouse incisor, *Eur J Oral Sci 118*, 118–130.
4. Schena, M., Shalon, D., Davis, R. W., and Brown, P. O. (1995) Quantitative monitoring of gene expression patterns with a complementary DNA microarray, *Science 270*, 467–470.
5. Watson, A., Mazumder, A., Stewart, M., and Balasubramanian, S. (1998) Technology for microarray analysis of gene expression, *Curr Opin Biotechnol 9*, 609–614.
6. Elvidge, G. (2006) Microarray expression technology: from start to finish, *Pharmacogenomics 7*, 123–134.
7. Mocellin, S., and Rossi, C. R. (2007) Principles of gene microarray data analysis, *Adv Exp Med Biol 593*, 19–30.
8. Roberts, P. C. (2008) Gene expression microarray data analysis demystified, *Biotechnol Annu Rev 14*, 29–61.
9. Wolf, B., Lesnaw, J. A., and Reichmann, M. E. (1970) A mechanism of the irreversible inactivation of bovine pancreatic ribonuclease by diethylpyrocarbonate. A general reaction of diethylpyrocarbonate. A general reaction of diethylpyrocarbonate with proteins, *Eur J Biochem 13*, 519–525.
10. Liu, H., Bebu, I., and Li, X. (2010) Microarray probes and probe sets, *Front Biosci (Elite Ed) 2*, 325–338.
11. Yang, I. V. (2006) Use of external controls in microarray experiments, *Methods Enzymol 411*, 50–63.
12. Yang, Y. H., Dudoit, S., Luu, P., Lin, D. M., Peng, V., Ngai, J., and Speed, T. P. (2002) Normalization for cDNA microarray data: a robust composite method addressing single and multiple slide systematic variation, *Nucleic Acids Res 30*, e15.
13. Churchill, G. A. (2004) Using ANOVA to analyze microarray data, *Biotechniques 37*, 173–175, 177.
14. Stekel, D. (2003) *Microarray Bioinformatics*, Cambridge University Press, Cambridge.
15. Benjamini, Y., Hochberg Y. (1995) Controlling the false discovery rate; a practical and powerful apporach to multiple testing., *J. R. Stat Soc B 57*, 589–300.
16. Pawitan, Y., Michiels, S., Koscielny, S., Gusnanto, A., and Ploner, A. (2005) False discovery rate, sensitivity and sample size for microarray studies, *Bioinformatics 21*, 3017–3024.
17. Pfaffl, M. W., Horgan, G. W., and Dempfle, L. (2002) Relative expression software tool (REST) for group-wise comparison and statistical analysis of relative expression results in real-time PCR, *Nucleic Acids Res 30*, e36.

Chapter 11

Lineage Differentiation of Mesenchymal Stem Cells from Dental Pulp, Apical Papilla, and Periodontal Ligament

Kentaro Akiyama, Chider Chen, Stan Gronthos, and Songtao Shi

Abstract

Recently, a variety of mesenchymal stem cells (MSCs), including dental pulp stem cells, stem cells from human exfoliated deciduous teeth, stem cells from apical papilla, periodontal ligament stem cells, and mesenchymal stem cells derived from human gingival, were isolated from orofacial and dental tissues. However, it is unknown whether these orofacial stem cells are derived from mesoderm or neural crest cell. In order to encourage orofacial MSC investigation, we provide detailed protocols for assessing lineage differentiation of orofacial MSCs.

Key words: Dental pulp stem cells, Stem cells from human exfoliated deciduous teeth, Stem cell from apical papilla, Periodontal ligament stem cells, Mesenchymal stem cells, Lineage differentiation

1. Introduction

Mesenchymal stem cells (MSCs) are non-hematopoietic multipotent stem cells capable of differentiating into both mesenchymal and non-mesenchymal cell types, including osteoblasts, adipocytes, and chondrocytes (1–5). Recently, multipotent MSCs were isolated from a variety of orofacial tissues, such as dental pulp-derived Dental Pulp Stem Cells (DPSCs (6)), Stem cells from Human Exfoliated Deciduous teeth (SHED (7)), periodontal ligament-derived Periodontal Ligament Stem Cells (PDLSCs (8)), root apical papilla-derived Stem Cell from Apical Papilla (SCAP (9)), dental follicle-derived progenitors from dental follicle (10), and gingival tissue-derived Gingival MSC (GMSC (11)). These orofacial MSCs showed significantly increased population doubling and proliferation rate compared to bone marrow-derived MSCs (6–9).

Chrissa Kioussi (ed.), *Odontogenesis: Methods and Protocols*, Methods in Molecular Biology, vol. 887,
DOI 10.1007/978-1-61779-860-3_11,

Thus, identification of these orofacial MSCs may provide unique cell source for tissue engineering and cell-based therapies. In this chapter, we provide experimental protocols for isolating and expanding DPSCs, SHED, PDLSCs, and SCAP and inducing specific lineage differentiation of these orofacial stem cells in vitro and in vivo.

2. Materials

2.1. Isolation and Primly Culture of DPSC, SHED, SCAP, and PDLSC

1. Alpha minimum essential medium (α-MEM).
2. Antibiotics, 100 U/mL penicillin and 100 μg/mL streptomycin.
3. 50-mL centrifuge tube.
4. Phosphate Buffered Saline (PBS), pH 7.4.
5. Digestion enzyme solution: 4 mg/mL Dispase II and 2 mg/mL Collagenase type I in PBS.
6. 100-mm Petri dish.
7. Povidone-iodide solution (10%w/v).
8. Sterilized gauze.
9. Periodontal scaler.
10. Scalpel with a blade.
11. Forceps.
12. Tooth extraction pliers.
13. Cutting pliers.
14. 100-mm cell strainer.
15. Complete growth medium; α-MEM supplemented with 15% (v/v) Fetal Bovine Serum (FBS), 2 mM L-glutamine, 0.1 mM L-ascorbic acid phosphate, and 100 U/mL of penicillin and 100 μg/mL streptomycin.
16. Trypan blue cell staining solution.
17. T 75 Culture flask.

2.2. Osteo/Odontogenic Differentiation In Vitro

1. Osteo/odonto-induction medium: α-MEM with 15% (v/v) FBS, 2 mM L-glutamine, 0.1 mM L-ascorbic acid phosphate, 100 U/mL of penicillin and 100 μg/mL streptomycin, 10^{-8} M dexamethasone, 1.8 mM KH_2PO_4.
2. 0.22-μm Filter unit.
3. Trypsin/EDTA solution.
4. 100-mm culture dish.
5. 60-mm culture dish.
6. 35-mm culture dish.

7. Cell scraper.
8. Tryzol regent.
9. 1.5-mL microcentrifuge tube.
10. Human PCR primer pairs: Human runx2 (GenBank accession no. L40992, sense 5-CAGTTCCCAAGCATTTCATCC-3, antisense 5-TCAATATGGTCGCCAAACAG-3), human osteocalcin (GenBank accession no. X53698, sense 5-CATGAGAG CCCTCACA-3, antisense 5-AGAGCGAC ACCCTAGA C-3), human BSP (GenBank accession no. L24759, sense 5-CTAT GGAGAGGACGCCAC GCCTGG-3, and antisense 5-CATAG CCATCGTAGCCTTGTCCT-3), human dentin sialophosphoprotein (DSPP) (sense 5-GGCAGTGACTCAAAAGGAGC-3, antisense 5-TGCTGT CACTGTCACTGCTG-3), and human GAPDH (GenBank accession no. M33197, sense 5-AGCC GCATCTTCTTTTGC GTC-3, antisense 5-TCATATTTGG CAGGTTTTTCT-3).
11. M-PER Mammalian protein extraction reagent.
12. Complete EDTA tablet.
13. Antibodies: Anti-Runx2 (Cat# PC287, 1:500 dilution, Calbichem), anti-ALP (Cat# sc-28904, 1:500 dilution, Santa Cruz), anti-OCN (Cat# AB10911, 1:500 dilution, Millipore), anti-BSP, anti-DSPP (From Dr. Larry Fisher, Craniofacial and Skeletal Disease Branch, NIDCR/NIH), and anti-β-actin (Cat#, 1:10,000 dilution, Sigma).
14. Alizarin Red S solution: 1% Alizarin red S in distilled water.
15. 60% (v/v) Isopropyl alcohol.

2.3. Lineage Differentiation In Vivo (Subcutaneous Transplantation in Mice)

1. Hydroxyapatite/tricalcium phosphate (HA/TCP) ceramic particles.
2. Immunocompromised mice; Beige nude XID III.
3. Anesthetic solution: Ketamine/xylazine.
4. 70% Alcohol gauze.
5. Povidone-iodide solution (10%w/v) swab.
6. Surgical scissors.
7. Surgical forceps.
8. Surgical suture.
9. Animal glue.

2.4. Regeneration of Calvarial Bone Defect

1. Immunocompromised mice.
2. HA/TCP ceramic particles.
3. Anesthetic solution: Ketamine/xylazine.
4. Surgical scissors.

5. Surgical forceps.
6. Surgical hand piece and motor with 1-mm-diameter round bur.
7. Surgical suture.
8. Animal glue.

2.5. Functional Bio Root Regeneration in Swine

1. Minipig.
2. Root-shape HA/TCP ceramic particles.
3. Gelform.
4. PDLSC and SCAP.
5. Tooth extraction elevator.
6. Surgical forceps.
7. Surgical suture.
8. Anesthetic solution: Ketamine/xylazine.

2.6. Recovery of Transplants

1. 4% Paraformaldehyde in PBS.
2. 10% Ethylenediaminetetraacetic acid (EDTA) in deionized water.
3. Ethanol.
4. Xylene.
5. Paraffin wax.
6. Hematoxylin solution.
7. Eosin solution.
8. 0.4% Acetic acid solution in distilled water.
9. Bluing solution.
10. Forceps.
11. Scalpel handle size 4.
12. Surgical blade size 10.
13. 30% Hydrogen peroxide.
14. Sodium azide.
15. Goat serum.
16. Antibodies.
17. Mounting medium.

3. Methods

3.1. Isolation of DPSC, SHED, SCAP, and PDLSC

Following informed consent, extracted teeth are collected from patients and stored in 50 mL α-MEM with antibiotics (penicillin 100 U/mL, streptomycin 100 μg/mL) at 4°C up to 24 h (see

Note 1). Third molars are recommended for isolation of DPSC and SCAP. Lower incisors from 6-year-old children are recommended for isolation of SHED (see Note 2). All procedure should be done in biohazard laminar flow hood and wear sterilized gloves to avoid contamination.

1. Aspirate medium and wash with autoclaved PBS three times, then transfer teeth onto 100-mm petri dish, and wipe surface with sterile gauze and 10% w/v povidone-iodine to remove debris.
2. For SCAP isolation, wash teeth with approximately 10 mL of autoclaved PBS three times, and then collect apical papilla tissue on the exterior of root foramen area using surgical blade.
3. For PDLSC isolation, remove soft tissue from cement–enamel junction (CEJ) with scalpel blade, and peal periodontal ligament from the surface of root using periodontal scaler or surgical blade. Transfer apical papilla tissue and periodontal ligament tissue into another clean petri dish with small amount of PBS (see Note 3).
4. For DPSC and SHED isolation, wash teeth by approximately 10 mL of PBS three times, then hold tooth with tooth extraction pliers, and split the tooth using cutting pliers at the level of CEJ. This step is not necessary in case root formation was not completed in permanent teeth or root absorption occurred in deciduous teeth. Next, hold split tooth with tooth extraction pliers, pull out pulp tissue using periodontal curettes or endodontic file, and transfer extracted pulp into another petri dish with small amount of PBS.
5. After cleaning and collecting tissues, cut isolated tissues into small pieces (see Note 4) with scalpel blade. Transfer into 50-mL centrifuge tubes with 5 mL of pre-warmed (37°C) digestion enzyme solution and incubate tubes for 60 min in 37°C water bath. Vortex tubes every 10 min to completely break up tissue.
6. After incubation, add 3 mL of growth medium to inactivate digestion enzymes and pass through a 100-mm cell strainer to get single suspension cell, then centrifuge tubes at $300 \times g$ for 10 min, and resuspend cells by 1 mL of complete growth medium.
7. Dilute 10 μL of cell suspension by 90 μL of 0.4% trypan blue cell staining solution to assess cell viability by using hematocytometer.
8. Prepare $2–3 \times 10^6$ cells into T 75 flasks for 3 h, and then wash culture flasks gently with PBS three times to eliminate unattached cells.

9. Put 10 mL of fresh growth medium into the flasks, change growth medium 7 days after seeding, and then culture further 7 days.
10. Approximately 10–14 days (see Note 5) after seeding cells, cells grow as a small cluster and form colonies.

3.2. Expansion of DPSC, SHED, SCAP, and PDLSC

Following dental tissues' harvest and stem cells' isolation, we routinely use these cells until five passages. The standard protocol for dental stem cells' subculture is necessary for keeping the cell quality for further experiments.

1. Wash ex vivo-expanded primary-cultured cells with 3 mL of PBS twice, and add 2 mL of trypsin/EDTA solution to digest cells in a 100-mm culture dish (or T75 flask) for 5 min at 37°C.
2. Use 2 mL of growth medium to inactivate trypsin/EDTA solution and transfer into a 50-mL centrifuge tube by pipetting, then centrifuge for 5 min at 300 × *g* to spin down cells, aspirate supernatant, and resuspend cells by growth medium.
3. Followed by count cell number as described above, seed 0.5×10^6 cells into a 100-mm culture dish or T75 flask for further expansion. Cells can be passed five times from primary culture (see Note 6).

3.3. Osteo/Odontogenic Differentiation of DPSC, SHED, SCAP, and PDLSC In Vitro

The ability of dental stem cells differentiates into osteo/odontogenic lineage in vitro and identifies lineage gene expression by reverse transcriptase polymerase chain reaction (RT-PCR) and Western blot is the standard procedure for stem cell characterization. The chemical staining, in terms of Alizarin red S staining, is the mineral module characterization standard procedure for osteo/odontogenesis.

3.3.1. Osteo/Odontogenesis Induction

1. Seed expanded cells into a 60-mm culture dish or a 35-mm culture dish at a density of 0.2×10^6 or 0.1×10^6, respectively, with growth medium, and culture cells at 37°C in 5% CO_2 until 100% confluent. Change medium twice a week.
2. Once cells reach to 100% confluence, change medium into osteo/odontogenic induction medium. Change induction medium twice a week.
3. Total RNA and protein can be harvested after a 1-week induction for RT-PCR and Western blot, and Alizarin red S staining can be performed to detect mineral deposition after 4 weeks' induction.

3.3.2. Reverse Transcriptase Polymerase Chain Reaction Analysis

1. Harvest osteo/odontogenic-induced cells using cell scraper and Trizol regent.
2. 100 ng of total RNA can be used for reverse transcription with super script reverse transcriptase III (Invitrogen Corporation).

3. Osteo/odontogenic gene expression can be amplified using specific primer pairs, including human runx2, human OCN, human BSP, human DSPP, and human GAPDH.

3.3.3. Western Blotting Analysis

1. Aspirate culture medium, wash with 3 mL of PBS twice, then add M-PER Mammalian protein extraction reagent with proteinase inhibitor, and measure protein concentration.
2. 10–20 μg of protein can be used for Western blotting analysis.
3. Osteo/odontogenic protein expression can be detected using specific antibody, including anti-Runx2, anti-ALP, anti-OCN, and anti-DSPP. Anti-β-actin can be used for loading control.

3.3.4. Alizarin Red S Staining

1. After 4–5 weeks' induction (see Note 7), aspirate induction medium and wash with PBS twice followed by fixing cells with 60% isopropyl alcohol for 1 min at room temperature.
2. Wash cells with distilled water for 2–3 min, and then stain with 1% Alizarin Red S staining solution for 10–15 min until mineralized deposits turn into red color at room temperature.
3. Wash cells with distilled water five times, then dry up, and observe mineralization under microscope.

3.4. In Vivo Osteo/Odontogenic Differentiation

For the purpose of examining the osteo/odontogenesis functions of ex vivo-expanded craniofacial stem cells in vivo, cultured cells are mixed with osteoconductive HA/TCP ceramic carrier particles followed by subcutaneous transplantation into immunocompromised mice.

1. When ex vivo-expanded cells grow to 120% confluent condition (see Note 8), prepare single cell suspension using trypsin/EDTA digestion and assess cell viability by using trypan blue.
2. Transfer 40 mg HA/TCP carrier particles to a 1.8-mL cryotube, mix gently 4×10^6 ex vivo-expanded cells by rotating, and incubate at 37°C for 90 min to attach cells onto the particles.
3. Pellet the cell-attached particles at $300 \times g$ for 6 min and remove supernatant.
4. Following an appropriate amount of ketamine and xylazine anesthesia to 8–10-week-old immunocompromised mice (Beige nude/nude XID III, Harlan), perform a 1.5-cm skin incision. Insert blunt-ended curved scissors under the skin and gently detach the skin from the muscle layer to create subcutaneous pockets (see Note 9) on both flanks. Place transplants into each subcutaneous pocket and close the incision with suture.
5. Collect the transplants 8 weeks after transplantation and fix with 4% paraformaldehyde for 1 day. Decalcify the transplants in 10% EDTA and 5% sucrose solution with daily changes about 3 weeks, and then follow standard paraffin embedding and histological staining protocols.

3.5. Regeneration of Calvarial Bone Defect

To assess the tissue regeneration capacity of MSCs, critical size of calvarial bone defect can be used. Mesenchymal stem cell and HA/TCP as carrier mixture can be transplanted into bone defect.

3.5.1. Prepare Cell and HA/TCP Mixture

1. Digest 100% confluent cells in a 100-mm dish with 2 mL of Trypsin/EDTA solution for 5 min at 37°C. Add 2 mL of growth medium to inactivate enzyme activity and transfer cells into a 50-mL centrifuge tube.
2. Centrifuge at $300\times g$ for 5 min, and resuspend cells in 1 mL growth medium.
3. Assess cell viability; $4–5\times10^6$ cells are incubated with 40 mg of HA/TCP carrier particles in a 1.8-mL cryotube for 90 min at 37°C. Centrifuge at $300\times g$ for 5 min and discard supernatant.
4. Keep transplants on ice until performing transplantation.

3.5.2. Prepare Critical Size of Bone Defect and Transplantation

1. Give anesthesia to immunocompromised mice with general protocol and clean skin by 70% ethanol and 10% povidone-iodide solution.
2. Cut skin and remove periosteum for preparing critical size (5–6 mm diameter, see Note 10) of bone defect with hand piece.
3. Transplant cell/HA/TCP mixture into bone defect, and close incision with five to six simple sutures and animal glue.

3.5.3. Recovery of Transplants and Histological Analysis

1. Collect the transplants including whole calvarial bone 8 weeks after transplantation, fix in 4% paraformaldehyde for 1 day at 4°C, then wash by PBS three times, and decalcify transplants with calvarial bone for 2–3 weeks in 40 mL of 10% EDTA solution in a 50-mL centrifuge tube by daily medium change.
2. After decalcification until transplant/bone become soft enough, wash samples with 40 mL of PBS, and dehydrate samples through increasing ethanol solutions 50, 70, 90, and 100% for 15 min and xylene for 15 min twice. Add molten paraffin wax before embedding twice and incubate for 1 h.
3. Embed and prepare 6–7-μm sections. Deparaffinize sections in xylene for 3 min twice. Rehydrate through a decreasing ethanol solutions, 100, 95, 90, 70, and 50%, followed by distilled water.
4. Stain with hematoxylin solution for 4 min and wash in tap water and 0.4% acetic acid for 10 s.
5. Wash in tap water, put into Bluing solution, and wash in running tap water.
6. Counterstain with eosin for 1 min and dehydrate in 100% ethanol. Immerse in xylene for 30 s three times and mount slides using xylene based liquid mounting media.

3.5.4. Immunohistochemistry

1. Deparaffinize section in xylene for 5 min twice.
2. Rehydrate through a decreasing ethanol solutions, 100, 95, 90, 70, and 50%, followed by distilled water.
3. Endogenous peroxidase activity is blocked using 3% hydrogen peroxide diluted in methanol for 30 min.
4. Wash sections with PBS three times, 5 min each, and block with 5% goat serum for 30 min at RT.
5. Primary antibodies are diluted as described in step 2
6. Wash sections with PBS twice, 3 min each.
7. Incubate with HRP/Fab Polymer Conjugate.
8. Wash with PBS five times, 3 min each.
9. Incubate with DAB solution.
10. Wash with distilled water for 3 min.
11. Wash with PBS for 3 min.
12. Incubate slides with hematoxylin solution for 1 min.
13. Wash with PBS for 30 s.
14. Dehydrate in 100% ethanol and xylene.
15. Mount with mounting medium.

3.6. Functional Bio Root Regeneration in Swine

3.6.1. Preparation of Bio Root

1. Prepare 10×10^6 PDLSCs with 1.5×1.5-cm gelform in growth medium for 3 days prior to transplantation.
2. Prepare approximate 10×10^6 SCAP with root-shaped HA/TCP for 2 h.
3. Surround PDLSC/gelform mixture with SCAP/HA/TCP mixture and keep it on ice until transplantation.

3.6.2. Transplantation of Bio Root in Swine

1. Give anesthetic solution using a general protocol, extract minipig lower incisor using elevator, and clean up the inside of socket using bone curette.
2. Place root-shaped SCAP/HA/TCP surrounded with PDLSC/gelform into socket, and add a pre-created post-channel inside of the root shape of HA carrier.
3. Close incision with five to six simple sutures, and leave it for 3 months without loading.
4. The SCAP/HA/TCP–PDLSC/gelform implant and post-channel were reexposed, and a premade porcelain crown was cemented to the SCAP/HA/TCP–PDLSC/gelform structure.

4. Notes

1. Teeth sample can be stored at 4°C for 24 h without any significant reduction of cell viability.
2. Unfinished-root-formation teeth are required for SCAP isolation.
3. Drying up tissue sample causes significant reduction of cell viability.
4. It is important to mince tissue as small as possible to obtain greater number of cells.
5. Culture period is variable. Once the center of colony reach to high density or multiple layers, pass the cells for further ex vivo expansion.
6. After passage 6, proliferation and differentiation capacity of MSC will be going down.
7. Due to accumulation of extracellular matrix, cells are detached from dish easily with inadequate wash.
8. More than 100% confluent condition is required to obtain greater bone formation following in vivo transplantation.
9. Make subcutaneous pocket as large as possible to reduce skin pressure against transplants.
10. Critical size of bone defect is important for evaluation of bone regeneration by stem cell. Less than 5-mm bone defect will be healed spontaneously.

References

1. Bianco, P., Riminucci, M., Gronthos, S. and Robey, P.G. (2001) Bone marrow stromal stem cells: nature, biology, and potential applications, *Stem Cells* 19, 180–192.
2. Friedenstein, A.J., Chailakhyan, R.K., Latsinik, N.V., Panasyuk, A.F. and Keiliss-Borok, I.V. (1974) Stromal cells responsible for transferring the microenvironment of the hemopoietic tissues. Cloning *in vitro* and retransplantation *in vivo*, *Transplantation* 17, 331–340.
3. Owen, M. and Friedenstein, A.J. (1988) Stromal stem cells: marrow-derived osteogenic precursors, *Ciba Found Symp* 136, 42–60.
4. Pittenger, M.F., Mackay, A.M., Beck, S.C., Jaiswal, R.K., Douglas, R., Mosca, J.D., Moorman, M.A., Simonetti, D.W., Craig, S., and Marshak, D.R. (1999) Multilineage potential of adult human mesenchymal stem cells, *Science* 284, 143–147.
5. Prockop, D.J. (1997) Marrow stromal cells as stem cells for nonhematopoietic tissues, *Science* 276, 71–74.
6. Gronthos, S., Mankani, M., Brahim, J., Robey, P.G., and Shi, S. (2000) Postnatal human dental pulp stem cells (DPSCs) in vitro and in vivo, *Proc Natl Acad Sci USA* 97, 13625–13630.
7. Miura, M., Gronthos, S., Zhao, M., Lu, B., Fisher, L.W., Robey, P.G., and Shi, S. (2003) SHED: Stem cells from human exfoliated deciduous teeth, *Proc Natl Acad Sci USA* 100, 5807–5812.
8. Seo, B.M., Miura, M., Gronthos, S., Bartold, P.M., Batouli, S., Brahim, J., Young, M., Robey, P.G., Wang, C.Y., and Shi, S. (2004) Investigation of multipotent postnatal stem cells from human periodontal ligament, *Lancet* 364, 149–155.

9. Sonoyama, W., Liu, Y., Fang, D., Yamaza, T., Seo, B.M., Zhang, C., Liu, H., Gronthos, S., Wang, C.Y., Wang, S., Shi, S. (2006) Mesenchymal stem cell-mediated functional tooth regeneration in Swine, *PLoS One* 1:e79.
10. Morsczeck, C., Götz, W., Schierholz, J., Zeilhofer, F., Kühn, U., Möhl, C., Sippel, C., Hoffmann, K.H. (2005) Isolation of precursor cells (PCs) from human dental follicle of wisdom teeth, *Matrix Biol* 24,155–165.
11. Zhang, Q., Shi, S., Liu, Y., Uyanne, J., Shi, Y., Shi, S. and Le, A.D. (2009) Mesenchymal stem cell derived colitis from human gingiva are capable of immunomodulatory functions and ameliorate inflammation-related tissue destruction in experimental clitis, *J Immunol* 183, 7787–7798.

Chapter 12

In Vivo Transplantation and Tooth Repair

Shuhei Tsuchiya and Masaki J. Honda

Abstract

Cell scaffold-based tooth engineering research was started by 2000 at Forsyth Institute corroborated with Dr. Vacanti's team at Massachusetts General Hospital. The first work was published in 2002 in Journal of Dental Research, in which we particularly focused on cells from postnatal tooth because of its clinical application. However, making a functional tooth from postnatal cells is still ways away. Alternatively, we formulated a partial replacement of the tooth by engineering the root of the tooth. Here, we describe a new technique in which the root of the third molar is used to replace missing teeth.

Key words: Dental follicle, Dental pulp, Enamel organ, Postnatal cell, Tissue engineering, Tooth root

1. Introduction

Human teeth are replaced only once during childhood. Therefore, when permanent teeth are lost, artificial reconstructions are required. Tissue engineering and stem-cell biology are currently the key strategies for tissue replacement in dentistry. A part of tissue, such as dentin and bone, has been already under clinical application, but whole tooth has not been under animal test yet. The great difference between dentin and tooth is that the epithelial-mesenchymal interaction is required to generate the entire tooth but not the dentin (1).

So far, there are several techniques available to reproduce a functional tooth as an organ. Approaches are categorized according to cell sources, the embryonic and postnatal cells. The results from both cell types differ considerably (2). The embryonic tooth germ cells can give rise to a tooth with correct structure and shape (3–5), but its application is very limited due to ethical and legal issues. In contrast, although it is easy to obtain an adult tooth, the

Chrissa Kioussi (ed.), *Odontogenesis: Methods and Protocols,* Methods in Molecular Biology, vol. 887,
DOI 10.1007/978-1-61779-860-3_12, © Springer Science+Business Media, LLC 2012

tissue-engineered tooth from postnatal tooth cells revealed an irregular morphology (6–8).

Here, we describe the methods for tooth regeneration using postnatal tooth bud from a 6-month-old pig (Fig. 1) (8). This approach is based on the concept that the scaffold is helpful in supporting the differentiation of progenitor cells. However, although tissues are regenerated with structures containing dentin and enamel, most produced tissues exhibit a disorganized heterogeneous morphology (Fig. 2) (9). Almost 10 years later, a procedure to make a functional new tooth has not yet established, especially using human dental stem cells, and thus we conclude that postnatal tooth germ cells have already lost the essential factors required to regulate tooth morphology (2).

From the clinical point of view, the most important part of the tooth is a root, which is supported by the periodontium. A tooth crown alone cannot fulfill the function of a tooth if the tooth root is missing. Recently, we established a method for the partial replacement of a missing tooth by engineering the tooth root. This strategy generated a root-shaped dentin mass associated with the periodontium, including cementum and periodontal ligament tissue. Briefly, the third molar was harvested and enamel organ, dental pulp, and dental follicle were separated. Only dental follicle cells

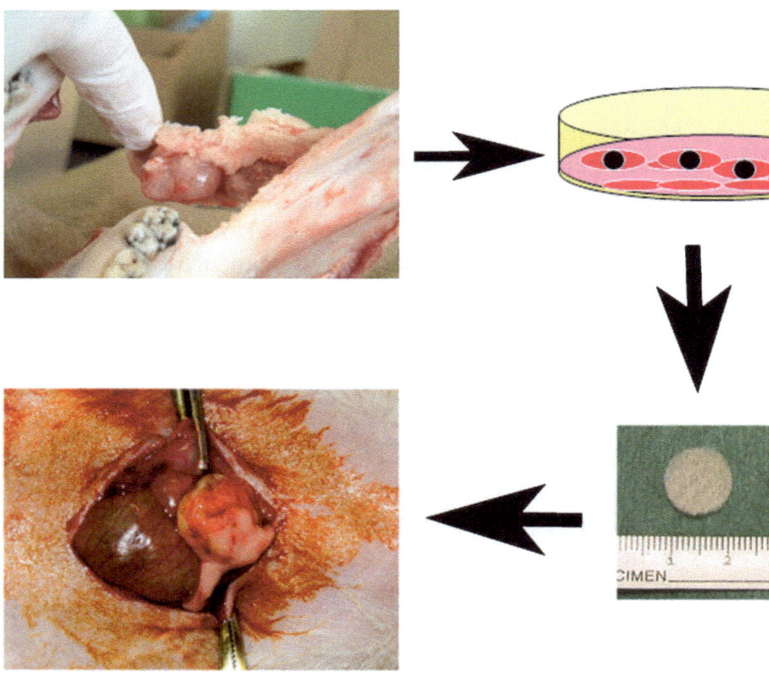

Fig. 1. Illustration of the strategy used to generate tooth by cell scaffold-based tissue engineering developed at Forsyth Institute. The third molar is harvested from the mandible. After elimination of mineralized tissue, tissues are dissociated into single-cell populations. Heterogeneous cell populations are seeded onto biodegradable polymer scaffold. The scaffold is transplanted into the omentum of nude rat.

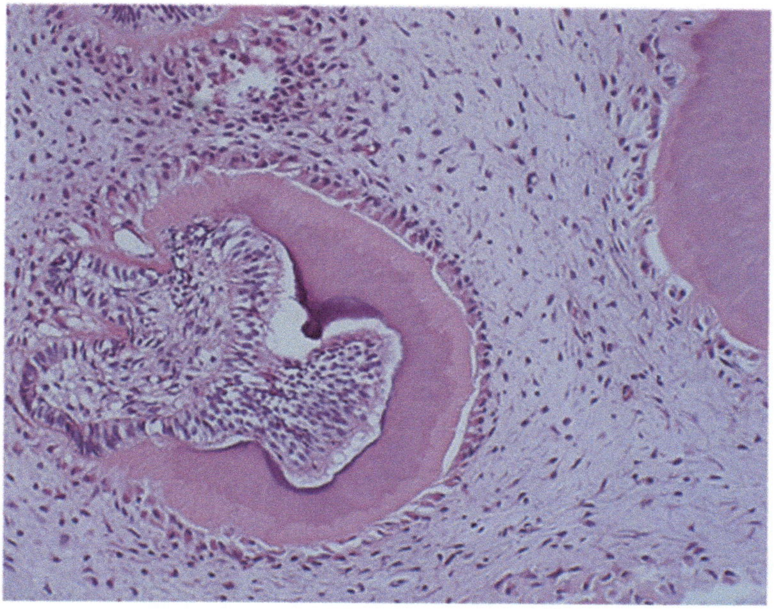

Fig. 2. Irregular shaped tooth tissues are formed in the scaffold after transplantation for 15 weeks.

were subcultured in vitro because of their paucity (10). Although the enamel organ culture was established (11–13), we were not able to expand dental pulp cells. Thus, we combined enamel organ, dental pulp, and subcultured dental follicle cells, which were created in the insert. Subcultured dental follicle cells were first seeded onto the bottom of the insert; and subsequently, dental pulp cells were placed on top of them, followed by enamel organ, and finally with a new layer of subcultured dental follicle cells. This approach mimics the tooth germ during its transplantion into the bone cavity of the mandible. The transplant was able to form dentin structures that surrounded dental pulp tissues filled with mineralized cementum or periodontal ligament-like tissues. In this chapter, we describe our new technique for making tooth root using tissue engineering technology.

2. Materials

All solutions are prepared with ultrapure water (prepared by purifying deionized water to a sensitivity of 18 MΩ cm at 25°C) and analytical grade reagents. Reagents are sterilized. Diligently follow all waste disposal regulations when disposing waste materials.

2.1. Cell Isolation and Culture Components

1. Cell culture medium: Dulbecco's modified Eagle's medium (DMEM) with 10% fetal bovine serum (FBS) and 1% penicillin. Store at 4°C (see Note 1).
2. Phosphate-buffered saline (PBS): Dulbecco's PBS 9.6 g into a 1-L glass bottle. Add 1 L of ultrapure water and sterilize with autoclave Store at 4°C (see Note 2).
3. Hank's balanced salt solutions (HBSS).
4. Pig mandible (see Note 3).
5. Collagenase solution: 2% Collagenase in HBSS (see Note 4).
6. 0.25% Trypsin–EDTA-4Na: 0.25% Trypsin, 1 mM EDTA-4Na in PBS (see Note 5).
7. Dispase II: 2,000 U Dispase II in HBSS.
8. 96 × 21-mm tissue culture dish.
9. 40-μm cell strainer (see Note 6).
10. Splitting osteotome.
11. Mallet.
12. Millex-GV (pore size 0.22 μm).
13. 6-, 12-, 24-, 48-, and 96-well culture dishes.
14. CO_2 incubator.
15. Micro-size constant temperature incubator shaker.
16. Hemocytometer.
17. Centrifuge.

2.2. Cylindrical Bone for Transplantation

1. Pig mandible (see Note 3).
2. Hand motor piece.
3. Dental bar instruments: Diamond point FG 1202, and Dentsply-Maillefer.

2.3. Construction of the Mimic Tooth Germ

1. Collagen gel: Rat tail collagen I.
2. Spatula: Microspatula.
3. Insert: 1.5-mL Eppendorf tube.

2.4. Implantation Components

1. Isodine.
2. Sterilized gauze.
3. 4-0 Vicryl.
4. 6-0 Nyron.
5. Size #15 Scalpel.
6. Surgical scissors.
7. Tweezers.
8. Pentobarbital.
9. F344/NJcl-*rnu/rnu* rat.

3. Methods

Carry out all procedures under sterile conditions. Autoclave all instruments. Cell culture media and PBS are pre-warmed at 37°C. Cells were cultured in 5% CO_2 at 37°C.

3.1. Cell Isolation and Culture

1. Fresh pig jaws obtained from the slaughterhouse and placed on ice during transportation (Fig. 3).
2. Split pig jaw in half, and remove muscle and connective tissue from bone using a razor blade (Fig. 4).
3. Break the jaw bone with small chisels (Fig. 5).
4. Obtain the third molar tooth germs under sterile conditions (Fig. 6) (see Note 7).
5. Place molars in 20 mL HBSS at room temperature in a 50-mL sterile conical tube.
6. Treat each tooth germ with 20 mL of 2,000 U Dispase II solution in a 50-mL centrifuge tube and agitate with Bioshaker for 1 h at 37°C (see Note 8).
7. Microdissect into dental follicle, enamel organ, and dental pulp from the tooth germ with scalpel under stereoscope (see Note 9).
8. Mince each tissue into about 2×2-mm square with scalpel. Add collagenase into culture dishes with minced tissues and transfer to 50-mL conical tubes.
9. Agitate with Bioshaker for 1 h at 37°C.
10. After enzymatic digestion, further dissociate the tissues by pipetting repeatedly using 10- and 5-mL pipettes for 10 min.

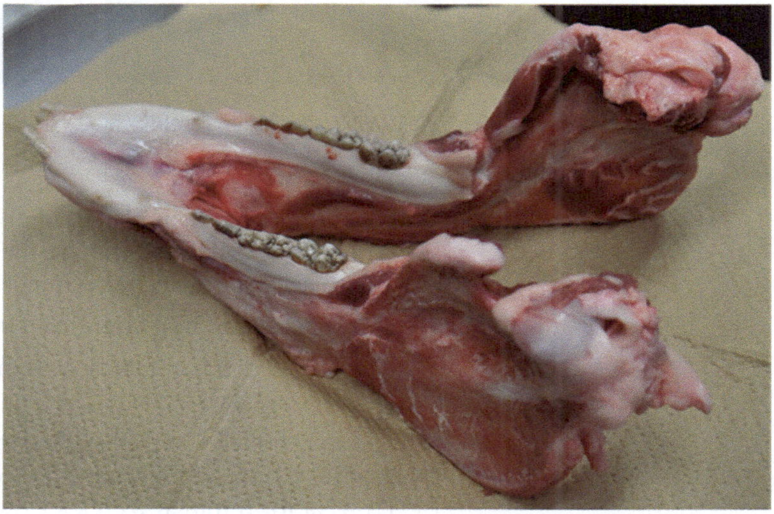

Fig. 3. Mandible of a 6-month-old pig.

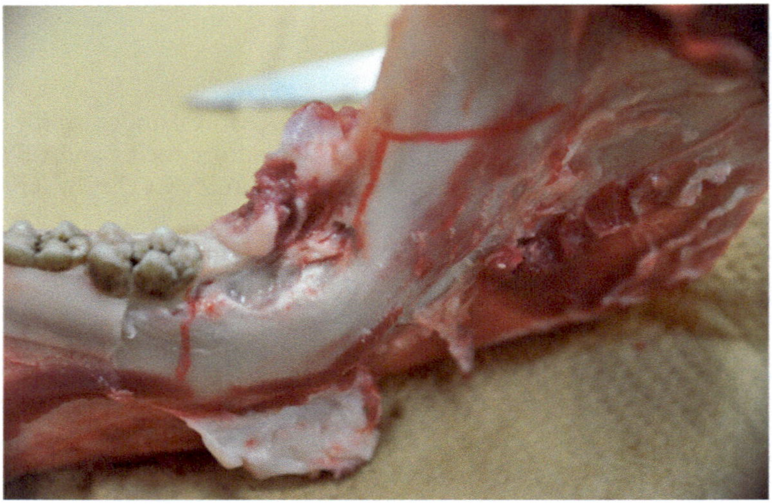

Fig. 4. Elimination of connective tissue and muscles from the surface of mandible.

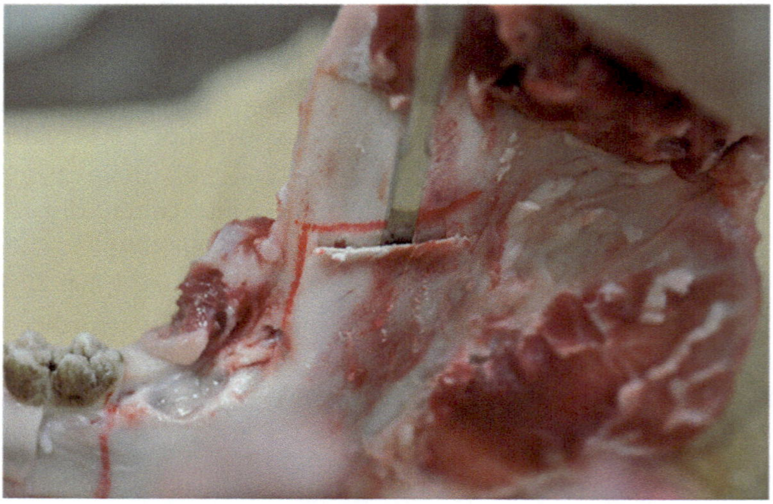

Fig. 5. Split of the mandible by chisel.

11. After washing once with HBSS, filter suspend with nylon filter (40-μm cell strainer) to eliminate the extracellular matrix, and centrifuge cells at $553 \times g$ for 5 min.
12. Discard supernatant and count cells with a hemocytometer.
13. Plate the isolated dental follicle cells on culture dish and culture them for several days.
14. After dental follicles are 70–80% confluent, wash them three times with 10 mL PBS. Dissociate cells with 0.05% Trypsin–EDTA-4Na for 5 min at 37°C.

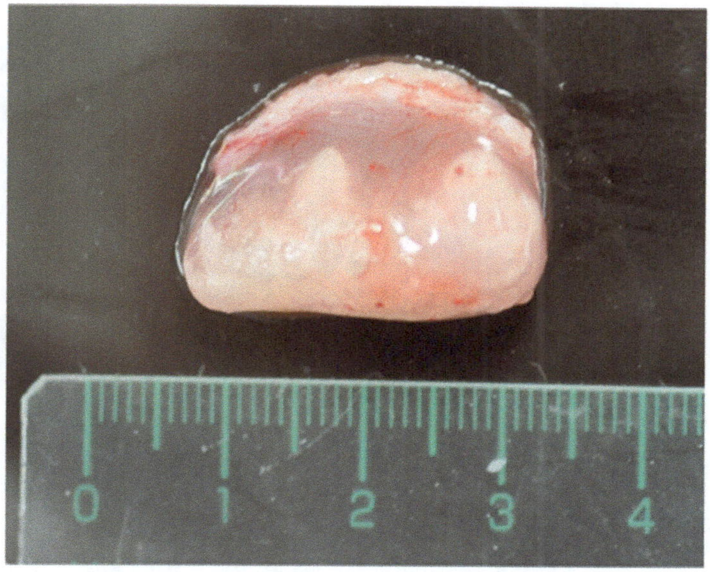

Fig. 6. Third molar of a 6-month-old pig. A small amount of the mineralized tissue is formed at this stage.

15. Observe whether the cells are detached from the culture dish with phase-contrast microscope. Add 10 mL of culture medium into the culture dish to prevent the effect of trypsin. Transfer the cell suspension to a 50-mL conical tube.
16. Count cells with hemocytometer, and resuspend them at a density of 1 cell/100 μL DMEM (see Note10).
17. Plate one cell/well of a 96-well plate and culture for 2 weeks. Replace medium every 3 days.
18. Single cell-derived dental follicles are transferred to a 6-well culture plate as in step 8 (see Note 11).
19. Dental follicle cells can be cultured until tenth passage.

3.2. Making the Cylindrical Bone Trunk

1. Excise mandibular ramus of the pig with hammer and chisel. Ramus has a cylindrical shape, 5 mm height, and 5 mm width. A 3-mm-diameter hole is made with dental air turbine by pouring PBS (Fig. 7) (see Note 12).
2. Wash three times the formed cylindrical bone trunk with HBSS to remove bone shavings.

3.3. Construction of Mimic Tooth Germ

1. Place the cylindrical bone trunk on a 6-well dish.
2. Obtain cultured dental follicle cells after trypsinization and centrifugation (see Note 13).
3. Count cells and dilute them 2.0×10^6 cells/50 μL collagen gel. Keep cells on ice (see Note 14).

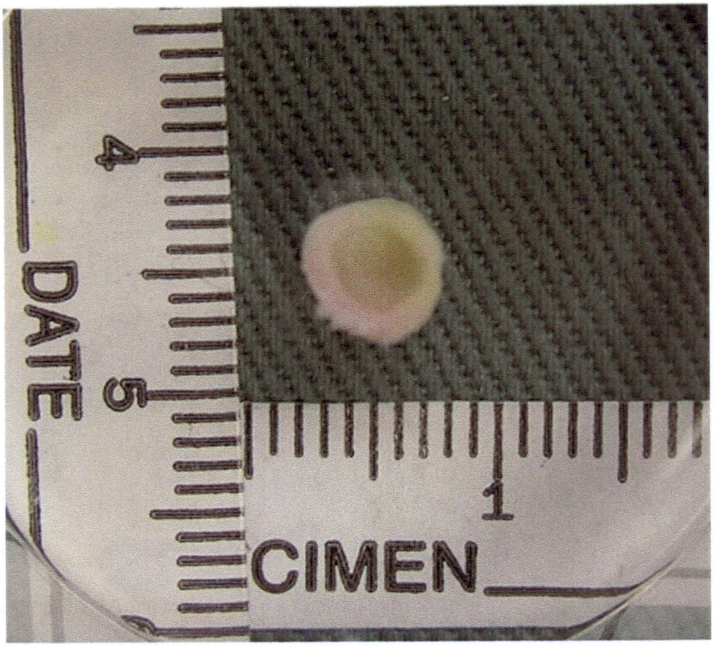

Fig. 7. Bone trunk from pig mandible.

4. Place carefully 25 μL of collagen gel containing 1.0×10^6 dental follicle cells into the bottom of the cylindrical bone trunk with micropipette. Incubate at 37°C, 5% CO_2, for 30 min (see Note 15).
5. Dilute both the 5.0×10^5 of enamel organ cells and 5.0×10^5 of dental pulp cells with 1.0 mL DMEM in an Eppendorf tube and centrifuge at $15{,}000 \times g$ for 20 min at 4°C. Filter cell pellets with a 23-gauge needle-attached 5-mL syringe, carefully. Spoon the floating cell pellet with a microspatula. Aspirate carefully the residual DMEM with a 23-gauge needle-attached 5-mL syringe. Place dental pulp cell pellet on top of the collagen gel contains dental follicle cells.
6. Place enamel organ cell pellet on top of dental pulp cell pellet.
7. Place the 25 μL of collagen gel containing dental follicle cells on the enamel organ pellet (Fig. 8) (see Note 16).
8. Incubate the bone trunk with the cell combination overnight at 37°C, 5% CO_2.

3.4. Transplantation

1. Anesthetize a 6-week-old F344/NJcl-*rnu*/*rnu* rat with diethyl ether for 2 min, followed by abdominal cavity injection of 50 mg/kg of nembutal. After 5 min, place rat on its back on cork plate folded with sterile drape and fixed legs with a 23-gauge needle (see Note 17).

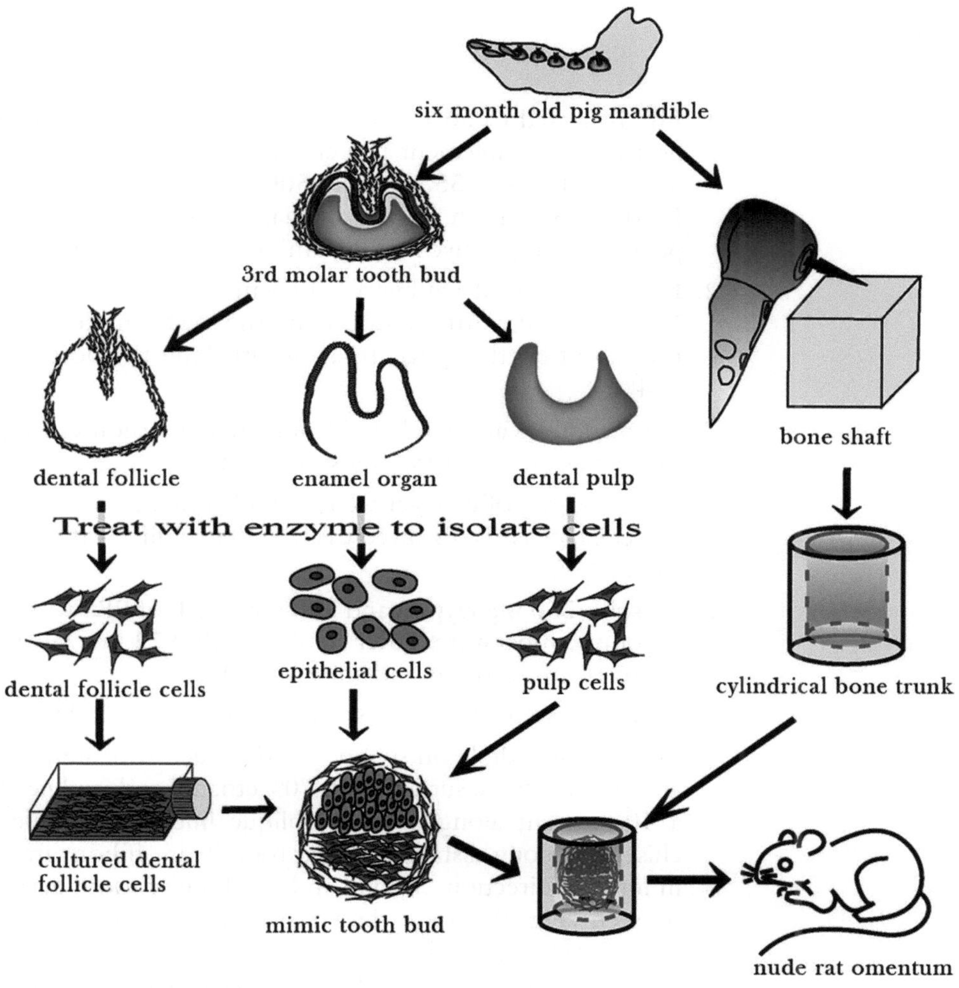

Fig. 8. Schematic diagram of our new approach for tooth root generation.

2. Sterilize the abdominal region with isodine. After ensuring the ensiform cartilage from surface of abdominal skin with tweezers, pick the skin and keep the tension; incise lower region about 20 mm.
3. Pick the liver with curette gently and confirm abdomen omentum under the liver. Spread the abdomen omentum with blunt tweezers.
4. Place the cell combination such as mimic tooth germ on the center of abdomen omentum, and fold with the rest of omentum. Fix the top of capsule is with 6-0 nylon suture (see Note 18).
5. Suture abdominal muscle with 4-0 vicryl and dermal skin with silk thread.

4. Notes

1. FBS is heated to 56°C for 30 min in a water bath to destroy heat-labile complement proteins prior to use in cell growth medium. Remove 55 mL from 500 mL in the plastic bottle of DMEM. Add both 50 mL of inactivated FBS and 5 mL of penicillin–streptomycin to 445 mL of DMEM. Store at 4°C.
2. Dissolve 9.6 g of Dulbecco's PBS in 1 L of ultrapure water. No need to stir. After autoclaving the medium, do not cover the cap of bottle tightly to avoid breaking the bottle. Store at 4°C.
3. Pig mandibles are purchased from a slaughterhouse. Keep the mandibles on ice during transportation.
4. Dilute 40 mg of collagenase in 20 mL HBSS. Vortex for 1 or 2 s; collagenase is dissolved easily. Filter sterilize with 0.22 μm Millex-GV.
5. Dissolve 250 mg trypsin and 380 mg in 1 L of PBS with stirrer at room temperature. Filter sterilize with 0.22 μm Millex-GV. Make 10-mL aliquots and store at −20°C.
6. When cell strainer is used, set it on a 50-mL conical tube.
7. Place pig mandible on styrene board folded with sterile paper, and sterilize bone surface with 70% ethanol and isodine. Make a 40-mm cut along internal oblique line of mandible with chisel and along distal first molar root. Tear off bone fragment in internal direction. Second and third molar tooth germs can be seen (Fig. 9). Since third molar tooth is connected with second molar teeth by dental lamina, dental lamina is cut from third molar with scalpel to detach from second molar. Thereafter, only third molar is pull out from tooth socket in

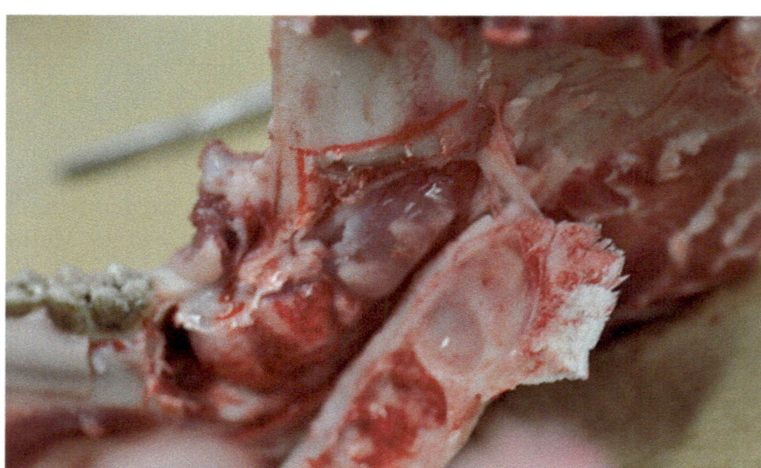

Fig. 9. Second and third molars are observed in mandible.

the mandible. Alter that, third molar teeth germ is only removed with sharp spoon. Keep tooth germs on ice.

8. Dispase II is pre-warmed at 37°C to facilitate the reaction. The agitation is intense, about 150 rpm/min.
9. After treating tooth germ with Diapase II, dental follicle is peeled from enamel organ spontaneously. Cut the dental follicle from tooth germ with scalpel. Enamel organ is separated from dental pulp with scalpel and tweezers. Discard mineralized cusps in tooth germs.
10. After pipetting, cell suspension is centrifuged at 1,500 rpm for 6 min. Thereafter, cells are diluted in 20 μL of HBSS and counted. Excess leftover cells are discarded.
11. When dental follicle cells grow to 70–80% confluence, aspirate medium and wash cells with PBS three times. Add 0.05% trypsin–EDTA-4Na to detach cells from plate. Incubate them at 37°C for 5 min. After confirming that cells are detached from the dish with phase-contrast microscopy, add medium with FBS to stop the reaction. Centrifuge the cell suspension at 1,500 rpm for 5 min, and aspirate the supernatant. Dilute cells into culture medium gently, and plate cells onto new plates in a desired density. When the cells are cultured from 96-well plate to 48-well plate, do not count cells to avoid loss.
12. Chisel pig mandibular ramus into a roughly 10×10-mm bone block. Place removed bone block in a sterile culture dish. Hold bone with tweezers and trimmed a square form with dental air turbine attached to Dentsply-maillefer bar. Round off the angle of square bone block and shape the bone block into cylinder form with diamond point FG 1202. Move the air turbine parallel to the bottom of dish when drilling the bone. The bone block will automatically form a cylinder because the tip of diamond point FG 1202 is flat. Finally, dig the center of the bone cylinder into 3-mm diameter and 3-mm depth with a diamond point FG 1202.
13. Usually, 10^6 dental follicle cells can be obtained from one culture dish.
14. When pipetting the cell suspension, try not to generate bubbles.
15. Touch the tip of pipet wall of bone cylinder in hole. By doing so, the shape of collagen gel in the cylinder makes a concave molding.
16. When treating cell pellet, do not steep in the medium. It makes difficult to treat the pellets. Treat the cell pellet with the tip of the 23-gauge needle.
17. Dilute Nembutal tenfold with PBS in a 1-mL syringe. Inject the diluted Nembutal slowly.
18. After suture, cut the suture thread short to avoid any stimulus that can be generated.

Acknowledgments

This work was supported in part by grants from the Japanese Ministry of Education, Culture, Sports, Science and Technology [Kakenhi KibanB (20659305) to MH], and Dental Research Center, Nihon University School of Dentistry.

References

1. Gronthos, S., M. Mankani, J. Brahim, P.G. Robey, and S. Shi. (2000). Postnatal human dental pulp stem cells (DPSCs) in vitro and in vivo. *Proc Natl Acad Sci USA.* 97:13625–30.
2. Honda, M.J., H. Fong, S. Iwatsuki, Y. Sumita, and M. Sarikaya. (2008). Tooth-forming potential in embryonic and postnatal tooth bud cells. *Med Mol Morphol.* 41:183–92.
3. Ikeda, E., R. Morita, K. Nakao, K. Ishida, T. Nakamura, T. Takano-Yamamoto, M. Ogawa, M. Mizuno, S. Kasugai, and T. Tsuji. (2009). Fully functional bioengineered tooth replacement as an organ replacement therapy. *Proc Natl Acad Sci U S A.* 106:13475–80.
4. Iwatsuki, S., M.J. Honda, H. Harada, and M. Ueda. (2006). Cell proliferation in teeth reconstructed from dispersed cells of embryonic tooth germs in a three-dimensional scaffold. *Eur J Oral Sci.* 114:310–7.
5. Nakao, K., R. Morita, Y. Saji, K. Ishida, Y. Tomita, M. Ogawa, M. Saitoh, Y. Tomooka, and T. Tsuji. (2007). The development of a bioengineered organ germ method. *Nat Methods.* 4:227–30.
6. Honda, M.J., Y. Sumita, H. Kagami, and M. Ueda. (2005). Histological and immunohistochemical studies of tissue engineered odontogenesis. *Arch Histol Cytol.* 68:89–101.
7. Sumita, Y., M.J. Honda, T. Ohara, S. Tsuchiya, H. Sagara, H. Kagami, and M. Ueda. (2006). Performance of collagen sponge as a 3-D scaffold for tooth-tissue engineering. *Biomaterials.* 27:3238–48.
8. Young, C.S., S. Terada, J.P. Vacanti, M. Honda, J.D. Bartlett, and P.C. Yelick. (2002). Tissue engineering of complex tooth structures on biodegradable polymer scaffolds. *J Dent Res.* 81:695–700.
9. Honda, M.J., Y. Shinohara, Y. Sumita, A. Tonomura, H. Kagami, and M. Ueda. (2006b). Shear stress facilitates tissue-engineered odontogenesis. *Bone.* 39:125–33.
10. Honda, M.J., M. Imaizumi, H. Suzuki, S. Ohshima, S. Tsuchiya, and K. Satomura. (2011). Stem cells isolated from human dental follicles have osteogenic potential. *Oral Surg Oral Med Oral Pathol Oral Radiol Endod.* 111:700–8.
11. Honda, M.J., T. Shimodaira, T. Ogaeri, Y. Shinohara, K. Hata, and M. Ueda. (2006a). A novel culture system for porcine odontogenic epithelial cells using a feeder layer. *Arch Oral Biol.* 51:282–90.
12. Honda, M.J., Y. Shinmura, and Y. Shinohara. (2009). Enamel tissue engineering using subcultured enamel organ epithelial cells in combination with dental pulp cells. *Cells Tissues Organs.* 189:261–7.
13. Honda, M.J., Y. Shinohara, K.I. Hata, and M. Ueda. (2007). Subcultured odontogenic epithelial cells in combination with dental mesenchymal cells produce enamel-dentin-like complex structures. *Cell Transplant.* 16:833–47.

Chapter 13

Methods to Validate Tooth-Supporting Regenerative Therapies

Miguel Padial-Molina, Julie T. Marchesan, Andrei D. Taut, Qiming Jin, William V. Giannobile, and Hector F. Rios

Abstract

In humans, microbially induced inflammatory periodontal diseases are the primary initiators that disrupt the functional and structural integrity of the periodontium (i.e., the alveolar bone, the periodontal ligament, and the cementum). The reestablishment of its original structure, properties, and function constitutes a significant challenge in the development of new therapies to regenerate tooth-supporting defects. Preclinical models represent an important in vivo tool to critically evaluate and analyze the key aspects of novel regenerative therapies, including (1) safety, (2) effectiveness, (3) practicality, and (4) functional and structural stability over time. Therefore, these models provide foundational data that supports the clinical validation and the development of novel innovative regenerative periodontal technologies. Steps are provided on the use of the root fenestration animal model for the proper evaluation of periodontal outcome measures using the following parameters: descriptive histology, histomorphometry, immunostaining techniques, three-dimensional imaging, electron microscopy, gene expression analyses, and safety assessments. These methods will prepare investigators and assist them in identifying the key end points that can then be adapted to later stage human clinical trials.

Key words: Guided tissue regeneration, Regenerative medicine, Bone regeneration, Tissue engineering, Animal models, Periodontal diseases, Periodontal engineering

1. Introduction

The tooth-supporting apparatus (i.e., periodontium) includes the alveolar bone, the periodontal ligament (PDL), the cementum, and the gingiva. Collectively, they represent a dynamic tissue complex with mechanical and biological functions that synergistically determine the tissue's adaptive potential and its ability to sustain microbiological and mechanical challenges. Through a number of

Chrissa Kioussi (ed.), *Odontogenesis: Methods and Protocols*, Methods in Molecular Biology, vol. 887, DOI 10.1007/978-1-61779-860-3_13, © Springer Science+Business Media, LLC 2012 complex mechanisms involving growth factors, transcription factors, and extracellular matrix (ECM) proteins, the periodontium is able to maintain its homeostasis, structure, and function and to respond and adapt to mechanical stimuli and infectious and/or inflammatory injuries (1, 2). However, once periodontal breakdown occurs, the ideal restoration (i.e., regeneration) of its original structure and function still remains a major challenge in the clinical setting (3). In general, these efforts have focused almost exclusively on regenerating lost alveolar bone. However, by definition, regeneration of the lost periodontium involves the formation of all tooth-supporting structures, including new cementum, PDL, alveolar bone, and gingival tissue. Also, the appropriate PDL tissue orientation, fiber directionality, and integration to both cementum and alveolar bone are required. Appropriate mechanical loading would be essential for the development of highly organized functional PDL fibers (4). Because of this critical interfacial connection of the multi-tissue complex that determines its function and stability, the use of periodontal-engineered devices has emerged as a prospective alternative to conventional treatments (5). Periodontal engineering uses life science and engineering technologies to restore the structure and function of alveolar bone, PDL, cementum, and gingival tissue (4).

The regenerative potential and the plausible biological mechanism of a novel therapy are often determined in vitro. However, to test the clinical feasibility and applicability of new therapies, the value of in vitro studies is very limited and frequently inadequate for direct entry into clinical trials (6). As required by regulatory approval agencies such as the US Food and Drug Administration and European Medicines Agency (EMEA), the safety and efficacy of new materials and techniques need to be tested in preclinical studies. Moreover, biological pathways taking place in these processes can also be studied and further validated (7). In addition, investigators can extrapolate their preclinical findings and identify important end points to be adapted to human clinical trial planning.

In general, the ideal preclinical model should be one that includes the following characteristics:

1. Standardization
2. Stable and controllable genetic background
3. Allow for evaluation of local and systemic safety
4. Facilitate the analysis of effectiveness by multiple modalities
5. Allow practical evaluation of functional and structural stability of the tissue over time
6. Cost-effective

Rodents (mice and rats) are the most commonly used animal models in biomedical research. Rats are cost-effective, easy to handle, and allow for the standardization of experimental conditions in genetically similar individuals (6). They are suitable for the study of the effects of physiological alterations related to aging, systemic diseases, pharmacological therapies, and immunodeficiency on tissue destruction and regeneration (8–10). Additionally, rat models allow evaluation of kinetics and biodistribution of different therapeutic agents, like adenovirus, by bioluminescence techniques. These techniques allow for safety evaluation and analysis of the short- and long-term biodistribution of the therapeutic agents administered both locally (11) and systemically (12) (Fig. 1). In vivo bioluminescence generated by expression of the luciferase transgene allows quantification and localization of transgene expression and provides noninvasive, dynamic, accurate, and comprehensive monitoring of vector expression systemically (12–14). Furthermore, analysis of luciferase-expressing recombinant

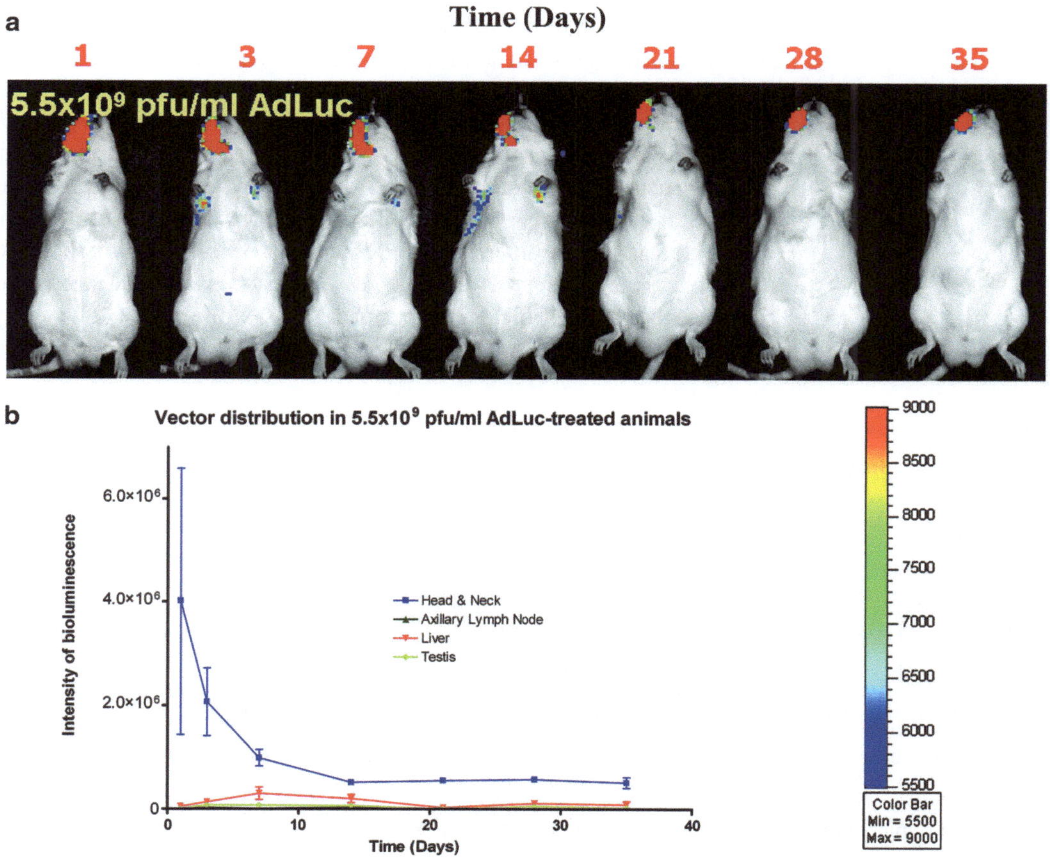

Fig. 1. The safety incorporation and processing of biologics are accurately monitored in the rat fenestration animal models as shown by the vector transduction efficiency and systemic distribution by bioluminescence. After the surgery on the right side, most of the luciferin signal is restricted to the alveolar bone defect region. However, a significant vector expression can be also noticed in distant organs, with maximum expression at day 14, followed by a decrease in vector expression in the head and neck region over time as well as in the maxillary area. Reproduced with permission from (12).

adenoviruses by bioluminescence is a highly sensitive method for evaluating the biodistribution and subsequent vector activity in the entire body (12).

To specifically study the regeneration of tooth-supporting structures, a widely used, accepted, and standardized model is required, for example an in vivo model that could assist on obtaining evidence for primary determination of the therapeutic efficacy, providing a proof of principle in a short time frame before proceeding to a larger animal model, etc. The rat fenestration model not only includes but also allows for the proper standardization of a number of these important aspects (Table 1). Based on this model, cementum and bone regeneration have been evaluated following the delivery of growth factors, genes, cells, and multiphasic scaffolds (5, 15–24). The extraoral approach of this model provides isolation from the oral environment, and thus can prevent negative effects, such as contamination, infection by intraoral microorganisms, or gingival tissue ingrowth.

For functional periodontal regeneration to occur, temporal and spatial progress in a similar sequence to that involved in the natural formation and development of the periodontium is needed (25). Although the exact cellular and molecular events are still not clear, cells must first migrate and attach to the denuded root surface. By using the rat fenestration defect model, a microenvironment that favors the proliferation, migration, and maturation of mesenchymal progenitors to the defect area of the PDL or the host bone has been observed (26, 27). This process is mediated and coordinated by soluble factors, other cells, and ECM. The early healing process follows the conserved sequence of wound healing

Table 1
Advantages and disadvantages of rat fenestration defect model

Advantages	Disadvantages
Proof of concept in a short time frame	Narrow healing time window
Well-contained defects	Small-size surgical microscopes required; technically challenging
No gingival tissue in growth	Rapid repair as kinetic healing model
Relatively low cost	Cannot measure healing of junctional epithelial-connective tissue interface
Controllable microflora	Not a "natural disease" model with microbial influence
Known age	Different anatomical structures compared to humans
Known genetic background	Different histopathologic features
Ease of handling and housing	Different host responses compared with humans

Adapted from refs. 6 and 30.

that is initiated by blood coagulation and migration of neutrophils and monocytes for wound debridement and bone resorption. Bone formation is typically initiated from the bony margins of the lesions (28). Within days after surgery, a thin cementum layer with a connective tissue attachment can be observed, particularly on the apical side of the teeth, where the cementum is thicker compared with the narrow coronal region (22). Once mineralized tissues are established, PDL fiber orientation, directionality, and integration to both cementum and alveolar bone are mediated by appropriate mechanical loading (4, 29). It is, therefore, crucial that investigators, according to the timeline that those processes follow (Fig. 2), select the appropriate time point(s) to determine the therapeutic efficacy "window" of a candidate periodontal-engineered device or bioactive molecule. In rats, recommended study evaluation time ranges from 2 to 6 weeks to capture early healing events and wound maturation (6).

Briefly, based on previously published procedures reviewed by Pellegrini et al. (6), Rios and Giannobile (30), and Seol et al. (31), we provide an overview of the rat fenestration model for the evaluation of periodontal outcome measures using descriptive histology, histomorphometry, immunostaining techniques, three-dimensional

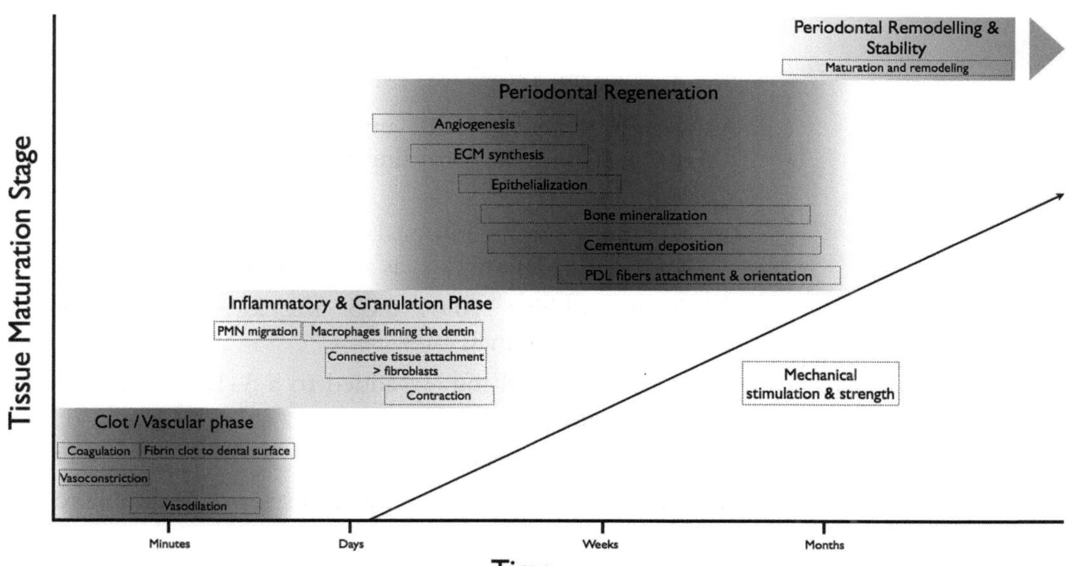

Fig. 2. Phases during periodontal healing and regeneration. Periodontal regeneration requires different processes in a sequential manner. After the initial coagulation phase, inflammatory reaction, and granulation tissue formation events, progenitor cells involved in multi-tissue regeneration are locally recruited and mediate the bioavailability of important growth factors. As the healing progresses, mechanical stimuli increase and promote an organized ECM synthesis as well as cementum and bone formation and maturation. Once those structures are established, PDL fibers are organized and oriented. Progressively, the tissues mature and ultimately increase their mechanical strength. Remodeling processes continue in the regenerated periodontium as an essential mechanism that monitors the adaptation potential to the challenging local and systemic environment.

imaging, electron microscopy, gene expression analyses, and safety assessments.

The procedure starts with an extraoral incision to reach the buccal aspect of the inferior molars. Buccal bone is removed up to the roots of the teeth to create a defect with standardized dimensions (e.g., $3 \times 2 \times 1$ mm). Roots are denuded, including the superficial dentin. Testing material can subsequently be applied to the defect and the flap repositioned and secured back in place.

2. Materials

1. Surroundings:
   (a) Sterile and sanitized surgical area.
      - Disinfectants, such as sodium hypochlorite, chlorine dioxide, dimethyl ammonium chloride, chlorine dioxide, or glutaraldehyde-based solutions.
      - Handwashing, sterile gloves, gowns, and masks.
      - Hot-bead instrument sterilizer, autoclave.
      - Hood and adequate ventilation system to assure aseptic conditions during the surgery.

   (b) Magnification/amplification system, such as a magnifying stereoscope (2–10×).
2. Anesthesia, analgesics, and antibiotics will vary with animal weight and must be according to veterinary instructions. In general:
   (a) Ketamine (intraperitoneal, IP, 40–90 mg/kg) and xylazine (IP, 5–10 mg/kg) as anesthetics.
   (b) Buprenorphine (0.01–0.05 mg/kg) and ketoprofen (5 mg/kg) as analgesics.
   (c) Ampicillin, 268 mg/L, added to a 5–10% dextrose solution as antibiotic.
   (d) 20–27-gauge needle.
3. Animal restraint and tissue retraction systems adaptable to animal size.
4. External heat source(s) (e.g., recirculating water blanket, microwaveable heating packs, or self-regulating heating pad).
5. Ophthalmic ointment (lubricant).
6. Povidone-iodine topical antiseptic, sterile saline, water, and/or 70% ethanol.
7. Hair removal blade, shaver, or cream.

8. Initial incision: Surgical blade (#11, 15), periosteal elevator (Pritchard).
9. Defect creation:
   (a) Surgical retractors, periodontal probe, 17/23 dental explorer.
   (b) Small, sharp hand instruments, such as Gracey curettes, hoes, or chisels.
   (c) Number ¼, and 4 round burs, low-speed and high-speed handpiece with engine and chisel.
10. Wound closure:
    (a) Needle holder (Crile-Wood).
    (b) Suture material (resorbable).
    (c) Scissors (LaGrange double curved).
    (d) Surgical clips (e.g., metal staples).
11. Sterile, clean cages for post-surgery animal recovery.
12. Staple remover.
13. Tissue harvesting: Scissors, round disc, and low-speed engine.
14. Tissue analyses:
    (a) Micro-computed tomography (CT) system for three-dimensional analysis of the mineralized tissues.
    (b) Histological, immunohistochemical, and immunofluorescence staining methods for specific molecules.
    (c) Optical microscope with imaging analysis apparatus. Ideally, a high-definition charge-coupled device (CCD) color camera capable of taking microscopic images is recommended.
    (d) Optical immunofluorescence microscope with imaging analysis apparatus or confocal microscope for capturing immunofluorescence images.
    (e) Histomorphometric and image analysis software.

3. Methods

3.1. Preparation for the Surgery

1. Acclimatization period of the animal of approximately 2 days to 1 week after arrival in a new housing facility.
2. Aseptic surgical area to perform an aseptic surgery (must not be used for any other purpose during the time of the surgery) (see Note 1).
3. Anesthesia with a combination of ketamine and xylazine via intraperitoneal injection, lasting for 45–90 min (see Note 2).

4. Apply ophthalmic ointment (lubricant) to the eyes of the animal to prevent drying.
5. Hair removal around the surgical area.
6. Skin disinfection with three alternating scrubs of povidone-iodine topical antiseptic, and warm, sterile saline water, or 70% ethanol (ethanol is less desirable) scrubbing in an outward and warm spiral direction.
7. Surgeons' preparation including, but not limited to, hand-washing, sterile gloves, gowns, and masks for each animal's surgical procedures.

3.2. Surgical Procedures

1. Animal restraint and retraction of soft tissues.
2. Surgery should be performed under a magnifying stereoscope (2–10×) to allow proper identification of anatomic landmarks and site preparation.
3. Identification of epithelial and hard tissue landmarks for the initial incision: parotid gland, masseter muscle, labial angle, inferior border of the mandible, molar teeth, etc.
4. First incision should be very superficial (only dermal layers) to expose the masseter muscle and gain access to a ligamentous landmark that extends in a postero-anterior direction approximating the lower border of the mandible (see Note 3).
5. Second incision is meant to dissect the area of interest through the masseter muscle slightly under the lower ligamentous line that could be visualized in an anterior–posterior direction until the body of the mandible is reached (buccal plate) (see Note 4).
6. Dissection of a distinct ligament that covers the area lateral to the first molar in order to ensure proper flap refection and adequate surgical access is required.
7. Once the bone is exposed and access to the first molar region is gained, the operator will be able to distinguish a more opaque and bulbous bone region with a tear-like shape, which is characteristic of the buccal plate.
8. The target defect creation area is the distal root of the mandibular first molar (buccal roots of the first and second molars can be included in the surgical defect). Initiate access with a no. 4 round bur, continue with a ¼ bur to complete the osteotomy, and remove the cementum once roots become visible. Standard bony defects should have 3×2×1 mm (see Note 5) (Fig. 3).
9. Apply the test agent(s) or regenerative device(s) into the created defect area according to the specific instruction that each material could require.
10. Reposition the muscle by using resorbable sutures.
11. Finally, reposition the skin by surgical clips.

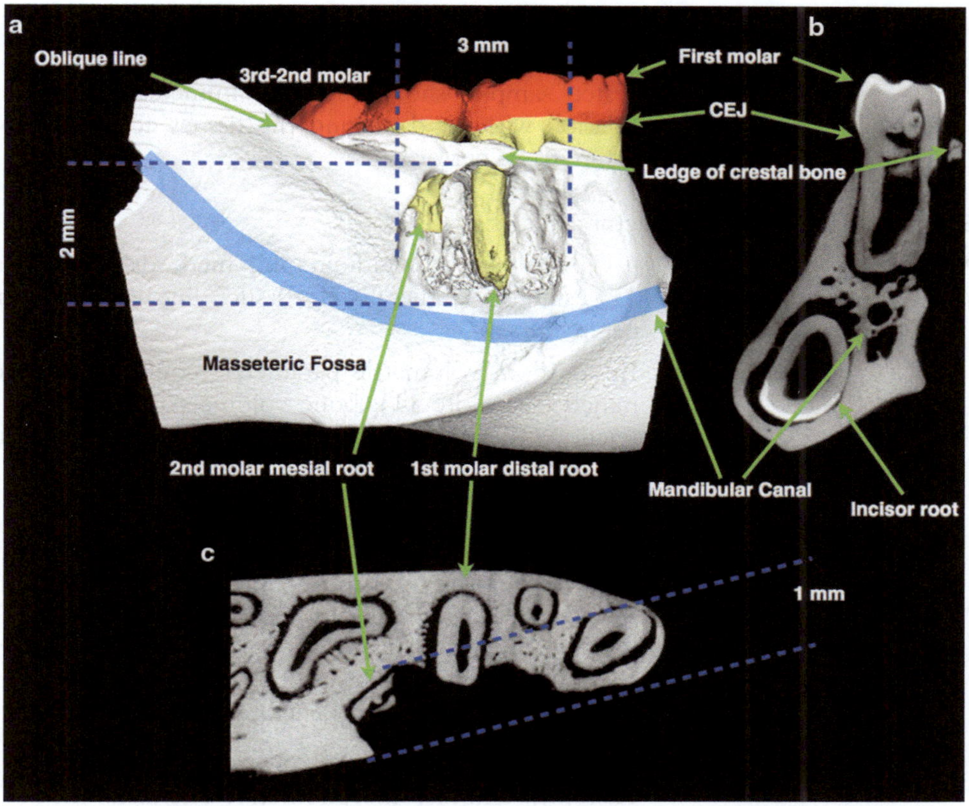

Fig. 3. μ-CT 3D reconstruction (**a**) and 2D sections (**b**: coronal; **c**: transversal) of a rat fenestration defect. Location, characteristics, and anatomical landmarks from different views are shown (18 × 18 × 18 μm^3 voxel size).

3.3. Postoperative Care

1. Analgesics should be administered for at least 24 h after the periodontal defect surgery (see Note 6).
2. Antibiotics dispensed via the water supply (see Note 7).
3. Animals should be mobile, fed freely following surgical recovery, and housed individually (see Note 8).
4. Reevaluation of the sutured wound at least three times per week is recommended until removal of clips at 2 weeks after surgery.
5. Treatments and monitoring must fit with the animal surgery guidelines given by the researcher's institution in accordance with regulations and in compliance with animal housing authorities (see Note 9).

3.4. Timing

1. At the designated end point, rats will be sacrificed using carbon dioxide overdose or according to institutional guidelines. A secondary method should be employed in order to confirm animal death prior to resuming tissue collection. This can be removal of vital organs.

2. Harvest tissues depending on the process desired to analyze and accordingly to the healing timeline (Fig. 2) (see Note 10).
3. Harvested samples should be fixed immediately to prevent degradation without damaging the tissues according to the procedure to be performed.

3.5. Endpoint Measurements

(a) Structural analysis:
- μ-CT evaluation or radiographic methods could be used to establish mineralized tissue lineal measurement parameters.
- In addition, volumetric parameters could also be determined (see Note 11): bone volume (BV), bone volume fraction (BVF), tissue mineral content (TMC), tissue mineral density (TMD), and bone mineral density (BMD).

(b) Biochemical analysis:
- Whole tissue dissection or laser capture microdissection (LCM) could be used to obtain tissue/cell samples from specific areas, such as PDL, bone, or cementum, for RNA or protein analysis (mRNA analysis, Western Blot, or ELISA techniques can be done to detect specific molecules relevant in periodontal regeneration).

(c) Cellular characterization:
- Histology and histomorphometry: The distal root of first molar is the main target for histologic and histomorphometric evaluation. PDL fibroblasts, osteoblast, cementoblasts, and fiber orientation from cementum to bone can also be studied (32).
- Immunohistochemical and immunofluorescence techniques allow the detection and immunolocalization of specific markers among the regenerated tissues.

4. Notes

1. Ideally, the surgical area can be located within the housing facilities, therefore limiting stress and potential health hazards to the animals. Disinfectants, such as sodium hypochlorite, chlorine dioxide, dimethyl ammonium chloride, or glutaraldehyde-based solutions, can be used to clean and disinfect the surgery area, although some may not be effective at eliminating all contaminants. Animals and instruments must also be prepared in a way to prevent contamination and ensure success of the survival surgery.

   All instruments should be cleaned and sterilized (e.g., autoclaved) prior to surgery. Disinfection/sterilization of multiple sets of instruments should be carried out for successive surgeries. Following use, instruments should be thoroughly cleaned before sterilization. Hot bead sterilization is a fast, dry method to prevent cross contamination between animals during surgery. Alternative sterilization methods may incorporate the use of glutaraldehyde or chlorine dioxide immersion followed by a sterile water or saline rinse. Aseptic techniques and sterile environments are critical to animal survival and positive experimental results.

Working in a laminar airflow hood can ensure the environment aseptic requirements.

2. Rat anesthetics and analgesics: A combination of ketamine (IP, 40–90 mg/kg) and xylazine (IP, 5–10 mg/kg) can be used as a general, injectable anesthesia for oral procedures. For prolonged anesthesia, supplement with one-third dose of ketamine only. IP injections should be performed using a 20–27-gauge needle that is inserted into the lower left abdominal quadrant with the animal in a head-down position. Anesthesia depth is typically monitored by the loss of response to external stimuli, such as a limb pinch.

   All animals should be provided an external heat source (e.g., recirculating water blanket, microwaveable heating packs, or self-regulating heating pad) in indirect contact with the animal to prevent hypothermia during the entire anesthesia and recovery period.

   Effective drug dosage may vary from animal to animal according to body weight, metabolism, and age. There are no exact calculations to relate the effective dose between animal and humans. Dosage can be determined by previous study results, published literature, and veterinary guidelines.

3. At this time, it is also important to separate the skin from the muscle around the incision line to allow proper closure and space for the use of surgical staples at the end of the surgery.

4. In some cases, the parotid gland duct (Stenson's) can be involved, causing postsurgical buccal swelling (mucoceles). This swelling can affect tissue regeneration as it produces mechanical pressure to the surgical area. Drainage of salivary secretions with a 10-mL syringe and 25-gauge needle may provide temporary relief from mechanical pressure. Parotid gland needs to be completely removed to eliminate swelling. Antibiotic water should be administered after removal of parotid gland.

5. During bone and cementum removal, it is difficult to irrigate with saline due to the small defect size, thin bone, and cementum. Special care should be taken not to generate heat damage at the surgical site, as it prevents a normal healing process. The PDL, cementum, and superficial dentin can be removed by a

combination of hand instrumentation and careful use of rotatory instruments. A very small ledge of crestal bone must remain coronally to maintain the integrity of the ridge and prevent communication with the oral cavity.

6. Buprenorphine (subcutaneous or intraperitoneal, 0.01–0.05 mg/kg) can be used for 8–12 h for postoperative pain relief, or for 24-h pain management, 5 mg/kg ketoprofen (subcutaneous) may be selected. Buprenorphine has negative interaction (which can lead to death of animal) with ketamine/xylazine cocktail if administered before animal recovers from anesthesia. So, if buprenorphine is used, it must be administered after rat awakens from anesthesia.
7. Ampicillin, 268 mg/L, added to a 5–10% dextrose solution can be used. Colored water bottles should be used with light-sensitive antibiotics.
8. Animal recovery time will vary with the type and dose of anesthesia, and may also vary between animals of similar sex, size, body mass, and genetic background.
9. For example, if biohazardous materials, such as viral vectors, are applied, the animal must be kept in biohazard facility until viral shedding has completed. Each virus will have a specific shedding period for which it can be considered contagious, and contact should be limited during this period. For example, adenovirus should be considered biohazardous for at least 72 h following application/inoculation.
10. By using this model, 3, 10, 21, and 35 days post-surgery time points are usually selected to harvest the tissue samples for analyzing bone and cementum regeneration.
11. Digital subtraction of teeth in the region of interest has to be done in order to exclude these structures from the analysis. Complete guidelines can be found in Park et al. (33).

Acknowledgments

The authors appreciate Sarah L. Volk for assisting with the preparation of the manuscript. We also acknowledge the technical contributions of Chan Ho Park, Scott Lim, and Jim Sugai. This work was supported by the NIH Grants K23DE019872 (HFR) and DE13397 (WVG). MPM was also supported by the Talentia Scholarship Program from the Regional Ministry for Innovation, Science and Enterprise, Junta de Andalusia (Spain). The authors report no conflicts of interest.

References

1. Bartold, P. M., and Narayanan, A. S. (2006) Molecular and cell biology of healthy and diseased periodontal tissues. *Periodontol 2000* 40, 29–49.
2. Rios, H. F., Ma, D., Xie, Y., Giannobile, W. V., Bonewald, L. F., Conway, S. J., and Feng, J. Q. (2008) Periostin is essential for the integrity and function of the periodontal ligament during occlusal loading in mice. *J Periodontol* 79, 1480–1490.
3. Bartold, P. M., Shi, S., and Gronthos, S. (2006) Stem cells and periodontal regeneration. *Periodontol 2000* 40, 164–172.
4. Rios, H. F., Lin, Z., Oh, B., Park, C. H., and Giannobile, W. V. (2011) Cell- and gene-based therapeutic strategies for periodontal regenerative medicine. *J Periodontol* 82, 1223–1237.
5. Park, C. H., Rios, H. F., Jin, Q., Bland, M. E., Flanagan, C. L., Hollister, S. J., and Giannobile, W. V. (2010) Biomimetic hybrid scaffolds for engineering human tooth-ligament interfaces. *Biomaterials* 31, 5945–5952.
6. Pellegrini, G., Seol, Y. J., Gruber, R., and Giannobile, W. V. (2009) Pre-clinical models for oral and periodontal reconstructive therapies. *J Dent Res* 88, 1065–1076.
7. Kaigler, D., Fuller, K., and Giannobile, W. V. (2010) Regulatory process for the evaluation of dental drugs, devices and biologics (Chapter 4). In: Giannobile, W. V., Burt, B. A., and Genco, R. J., Eds. Clinical Research in Oral Health 1 ed., Wiley-Blackwell, Ames, pp 55–78.
8. Benatti, B. B., Neto, J. B. C., Casati, M. Z., Sallum, E. A., Sallum, A. W., and Nociti Jr, F. H. (2006) Periodontal healing may be affected by aging: A histologic study in rats. *J Periodontal Res* 41, 329–333.
9. Graves, D. T., Fine, D., Teng, Y. T. A., Van Dyke, T. E., and Hajishengallis, G. (2008) The use of rodent models to investigate host-bacteria interactions related to periodontal diseases. *J Clin Periodontol* 35, 89–105.
10. Klausen, B. (1991) Microbiological and immunological aspects of experimental periodontal disease in rats: a review article. *J Periodontol* 62, 59–73.
11. Jin, Q., Anusaksathien, O., Webb, S. A., Printz, M. A., and Giannobile, W. V. (2004) Engineering of tooth-supporting structures by delivery of PDGF gene therapy vectors. *Mol Ther* 9, 519–526.
12. Chang, P. C., Cirelli, J. A., Jin, Q., Seol, Y. J., Sugai, J. V., D'Silva, N. J., Danciu, T. E., Chandler, L. A., Sosnowski, B. A., and Giannobile, W. V. (2009) Adenovirus encoding human platelet-derived growth factor-B delivered to alveolar bone defects exhibits safety and biodistribution profiles favorable for clinical use. *Hum Gene Ther* 20, 486–496.
13. Johnson, M., Huyn, S., Burton, J., Sato, M., and Wu, L. (2006) Differential biodistribution of adenoviral vector in vivo as monitored by bioluminescence imaging and quantitative polymerase chain reaction. *Hum Gene Ther* 17, 1262–1269.
14. Wood, M., Perrotte, P., Onishi, E., Harper, M. E., Dinney, C., Pagliaro, L., and Wilson, D. R. (1999) Biodistribution of an adenoviral vector carrying the luciferase reporter gene following intravesical or intravenous administration to a mouse. *Cancer Gene Ther* 6, 367–372.
15. Giannobile, W. V., Lee, C. S., Tomala, M. P., Tejeda, K. M., and Zhu, Z. (2001) Platelet-derived growth factor (PDGF) gene delivery for application in periodontal tissue engineering. *J Periodontol* 72, 815–823.
16. Howell, T. H., Fiorellini, J. P., Paquette, D. W., Offenbacher, S., Giannobile, W. V., and Lynch, S. E. (1997) A phase I/II clinical trial to evaluate a combination of recombinant human platelet-derived growth factor-BB and recombinant human insulin-like growth factor-I in patients with periodontal disease. *J Periodontol* 68, 1186–1193.
17. Huang, K. K., Shen, C., Chiang, C. Y., Hsieh, Y. D., and Fu, E. (2005) Effects of bone morphogenetic protein-6 on periodontal wound healing in a fenestration defect of rats. *J Periodontal Res* 40, 1–10.
18. Jin, Q. M., Anusaksathien, O., Webb, S. A., Rutherford, R. B., and Giannobile, W. V. (2003) Gene therapy of bone morphogenetic protein for periodontal tissue engineering. *J Periodontol* 74, 202–213.
19. Jin, Q. M., Zhao, M., Webb, S. A., Berry, J. E., Somerman, M. J., and Giannobile, W. V. (2003) Cementum engineering with three-dimensional polymer scaffolds. *J Biomed Mater Res A* 67, 54–60.
20. King, G. N., and Hughes, F. J. (1999) Effects of occlusal loading on ankylosis, bone, and cementum formation during bone morphogenetic protein-2-stimulated periodontal regeneration in vivo. *J Periodontol* 70, 1125–1135.
21. King, G. N., and Hughes, F. J. (2001) Bone morphogenetic protein-2 stimulates cell recruitment and cementogenesis during early wound healing. *J Clin Periodontol* 28, 465–475.

22. King, G. N., King, N., Cruchley, A. T., Wozney, J. M., and Hughes, F. J. (1997) Recombinant human bone morphogenetic protein-2 promotes wound healing in rat periodontal fenestration defects. *J Dent Res* 76, 1460–1470.
23. Talwar, R., Di Silvio, L., Hughes, F. J., and King, G. N. (2001) Effects of carrier release kinetics on bone morphogenetic protein-2-induced periodontal regeneration in vivo. *J Clin Periodontol* 28, 340–347.
24. Zhao, M., Jin, Q., Berry, J. E., Nociti Jr, F. H., Giannobile, W. V., and Somerman, M. J. (2004) Cementoblast delivery for periodontal tissue engineering. *J Periodontol* 75, 154–161.
25. Chen, F. M., An, Y., Zhang, R., and Zhang, M. (2011) New insights into and novel applications of release technology for periodontal reconstructive therapies. *J Control Release* 149, 92–110.
26. Lekic, P., Sodek, J., and McCulloch, C. A. G. (1996) Osteopontin and bone sialoprotein expression in regenerating rat periodontal ligament and alveolar bone. *Anat Rec* 244, 50–58.
27. Lekic, P., Sodek, J., and McCulloch, C. A. G. (1996) Relationship of cellular proliferation to expression of osteopontin and bone sialoprotein in regenerating rat periodontium. *Cell Tissue Res* 285, 491–500.
28. Rajshankar, D., McCulloch, C. A. G., Tenenbaum, H. C., and Lekic, P. C. (1998) Osteogenic inhibition by rat periodontal ligament cells: Modulation of bone morphogenic protein-7 activity in vivo. *Cell Tissue Res* 294, 475–483.
29. Mine, K., Kanno, Z., Muramoto, T., and Soma, K. (2005) Occlusal forces promote periodontal healing of transplanted teeth and prevent dentoalveolar ankylosis: an experimental study in rats. *Angle Orthod* 75, 637–644.
30. Rios, H. F., and Giannobile, W. V. (2011) Preclinical protocols for periodontal regeneration (Chapter 7). In: Osteology Guidelines for oral and maxillofacial regeneration: Preclinical models for translational research. Giannobile, W. V. and Nevins, M. Eds. Berlin: Quintessence Publishing Co, Inc, pp 77–104. ISBN: 9781850972112.
31. Seol, Y. J., Pellegrini, G., Franco, L. M., Chang, P. C., Park, C. H., and Giannobile, W. V. (2010) Preclinical methods for the evaluation of periodontal regeneration in vivo. *Methods Mol Biol* 666, 285–307.
32. Park, C. H., Rios, H. F., Jin, Q., Sugai, J. V., Padial-Molina, M., Taut, A. D., Flanagan, C. L., Hollister, S. J., and Giannobile, W. V. (2012) Tissue engineering bone-ligament complexes using fiber-guiding scaffolds. *Biomaterials* 33, 137–145.
33. Park, C. H., Abramson, Z. R., Taba Jr, M., Jin, Q., Chang, J., Kreider, J. M., Goldstein, S. A., and Giannobile, W. V. (2007) Threedimensional micro-computed tomographic imaging of alveolar bone in experimental bone loss or repair. *J Periodontol* 78, 273–281.

Chapter 14

Generation of a Bioengineered Tooth by Using a Three-Dimensional Cell Manipulation Method (Organ Germ Method)

Masamitsu Oshima, Miho Ogawa, Masato Yasukawa, and Takashi Tsuji

Abstract

The arrangement of cells within a tissue plays an essential role in organogenesis, including tooth development. Organ morphogenesis and physiological functions induced by three-dimensional tissue organization are well known to be regulated by the proper spatiotemporal organization of various signaling molecules, including cytokines, extracellular matrix proteins, and adhesion molecules. Development of a three-dimensional cell manipulation technology to create a bioengineered organ germ, designated as the organ germ method, enabled the generation of a structurally correct and fully functional bioengineered tooth in vivo. This method is expected to be utilized as a valuable technique for analyzing gene and protein functions during organogenesis. Here, we describe protocols for tooth germ reconstitution using the organ germ method and methods for analyzing tooth development in vitro and in vivo.

Key words: Bioengineered tooth, Organ germ method, Cell manipulation, Transplantation, Tooth germ

1. Introduction

The tooth is an ectodermal organ arising from the tooth germ, which is induced by reciprocal interactions between the oral epithelium and mesenchyme in the developing embryo (1–3). Following tooth germ formation, the epithelial and mesenchymal cells in the tooth germ differentiate into the cells, which form tooth tissue, including ameloblast, odontoblast, and pulp cells, as well as periodontal ligament cells (4, 5). These cells secrete the minerals that create hard tissues, such as enamel, dentin, cementum, and alveolar bone (4, 5). Various morphological features

Chrissa Kioussi (ed.), *Odontogenesis: Methods and Protocols*, Methods in Molecular Biology, vol. 887,
DOI 10.1007/978-1-61779-860-3_14, © Springer Science+Business Media, LLC 2012

define different types of teeth, such as incisors, canines, and molars, which are defined by the macromorphological features of crown size and tooth length and by the micromorphological features of the number and position of the cusp and roots (6, 7). Tooth development and morphogenesis have been elucidated to be regulated by the proper spatiotemporal expression of various signaling molecules, such as cytokines, extracellular matrix, and adhesion molecules, which are induced by epithelial and mesenchymal interactions (8).

Proper three-dimensional organization of cells within a tissue has established essential roles in governing varied aspects of organogenesis, including regulation of cell growth, differentiation, and functions (9–11). In research on tooth development, investigations into the molecular mechanisms of odontogenesis have been performed using in vivo experiments and/or in vitro organ culture, but have not utilized a two-dimensional cell culture (12). Furthermore, many attempts to regenerate a third tooth from bioengineered tooth germ, reconstituted from tooth germ-derived epithelial and mesenchymal cells using a three-dimensional cell manipulation technique such as cell aggregation and biodegradative scaffold methods, have also been conducted with the goal of developing methodologies for regenerative tooth therapy (13–15).

We have recently developed an in vitro three-dimensional cell manipulation method, designated as the bioengineered organ germ method, which incorporates cell compartmentalization between epithelial and mesenchymal cells at a high cell density (16). The bioengineered tooth germ can generate a structurally correct and fully functioning bioengineered tooth at high frequency after transplantation into the oral environment of an adult mammal in vivo (17). Gene expression profile analyses of the gene networks that regulate early tooth development revealed that our bioengineered tooth germ replicates the tooth organogenesis observed during embryonic development. Thus, our three-dimensional, bioengineered organ technique can be employed for use in various studies on gene networks during tooth development and screenings for potential tooth-inductive stem cells. This technique is also useful for investigations into three-dimensional morphogenesis and cell movement analysis during organ development processes achieved through the upregulation or downregulation of gene expression or protein functions at the single-cell level (7, 16, 17).

In this chapter, we provide a detailed description of a protocol for three-dimensional bioengineered tooth germ reconstitution using tooth germ-derived epithelial and mesenchymal cells. We also describe methods for analyses utilized for the in vitro and in vivo study of tooth development.

2. Materials

All solutions should be prepared using ultrapure water (prepared by purifying deionized water to attain a sensitivity of 18 MΩ cm at 25°C) and analytical grade reagents. All surgical instruments should be washed and sterilized in an autoclave prior to each use in order to prevent contamination. It is recommended that enzyme and serum reagents should be evaluated for enzyme reactivity and primary cell culture efficiencies, respectively, prior to experimental use.

1. Animals: An inbred mouse strain (i.e., C57BL/6, Balb/c 3T3, etc.) should be used in these experiments. Mouse embryonic age is determined based on the day of appearance of a vaginal plug in a pregnant mouse (embryonic day 0; ED 0).
2. Ca^{2+}/Mg^{2+}-free, phosphate-buffered saline (PBS(−)): 137 mM sodium chloride (NaCl), 2.7 mM potassium chloride (KCl), 8.0 mM anhydrous disodium hydrogen orthophosphate (Na_2HPO_4), and 1.5 mM potassium phosphate monobasic (KH_2PO_4). Store at 4°C (see Note 1).
3. Medium: Isolation of tooth germ and dissociation of single cells are performed in basal cell culture medium supplemented with 10% fetal bovine serum (FBS), 1% penicillin–streptomycin, and 10 mM 1-4-(2-hydroxyethyl)-piperazineethanesulfonic acid (HEPES). For organ culture, this medium is added with the above-mentioned supplements, excluding HEPES (see Note 2).
4. Dispase solution: Dispase (50 U/mL, Cat# 354235; BD, Franklin Lakes, NJ, USA) is diluted tenfold in Hanks' balanced salt solutions (HBSS) and stored at −20°C until use (see Note 3).
5. Collagenase solution: The concentration of collagenase I (Cat# 4196; Worthington, Lakewood, NJ, USA) is adjusted to 1 mg/mL using distilled water and stored at −20°C until use (see Note 4).
6. 2.5% Trypsin (tenfold concentrated).
7. Reagent A: Component in a collagenase solution consisting of 7.32 μL in 2 mL PBS(−).
8. Reagent B: Component in a trypsin solution consisting of 100 μL in 900 μL PBS(−).
9. Reagent C: Component in a solution consisting of 1.83 μL collagenase and 100 μL trypsin in 900 μL PBS(−).
10. 70 U/mL Deoxyribonclease I (DNase) from bovine pancreas (see Note 5).
11. Collagen gel: Component in 100 μL of tenfold concentrated α-Minimum Essential Medium (αMEM) (Sigma) and 100 μL

mixed buffer (0.08 N sodium hydroxide and 200 mM HEPES) in 800 μL Cellmatrix Type I-A (see Note 6).

12. Methylene blue gel (see Note 7).
13. Surgical tools for the dissection of embryos from the uterus: Large surgical scissors and forceps to cut the abdominal skin and muscle, and small surgical scissors and forceps to dissect the uterus (see Note 8).
14. Surgical tools for transplantation into jaw's bone: Disposable surgical knife to cut off the gingiva, and drills to prepare a hole in the jaw's bone for implantation (see Note 9).
15. Dissecting microscope (see Note 10).
16. Sterile disposable 1-mL syringes and 25-G needles (5/8; 0.50 × 16 mm) for tooth germ extraction (see Note 11).
17. Sterile disposable Petri dishes (35- and 100-mm).
18. Sterile disposable polypropylene conical tube (15-mL) to collect single cells isolated from tooth germ.
19. Sterile disposable 1.5-mL microtube (see Note 12).
20. Sterile disposable gel loader tips and pipette tips for reconstitution of bioengineered tooth germ (see Note 13).
21. Culture at the medium-gas interface (cell culture insert/0.4-μm pore size membrane, see Note 14).
22. Silicon grease.
23. Sterile micropipette (see Note 15).

3. Methods

Dissecting manipulation of embryos, tooth germs, and tooth tissues should be performed in medium to prevent drying.

3.1. Extraction of Mouse Embryo Tooth Germ

1. After sacrificing the mouse, cut the abdomen along the midline with small scissors. Resect the uterus and wash it with PBS(−) (see Note 16).
2. Dissect the embryos from the uterus. Amputate the fetal head from the body and separate the lower jaw from the head. Immediately place the lower jaw in cold (4°C) culture medium (see Note 17).
3. Resect the tongue from the lower jaw (Fig. 1a–c, see Note 18) and separate a unilateral jaw (Fig. 1d, e).
4. Resect Meckel's cartilage from the lower jaw (Fig. 1f, see Note 19) and then resect the tissue of the opposite side from Meckel's cartilage (Fig. 1g).

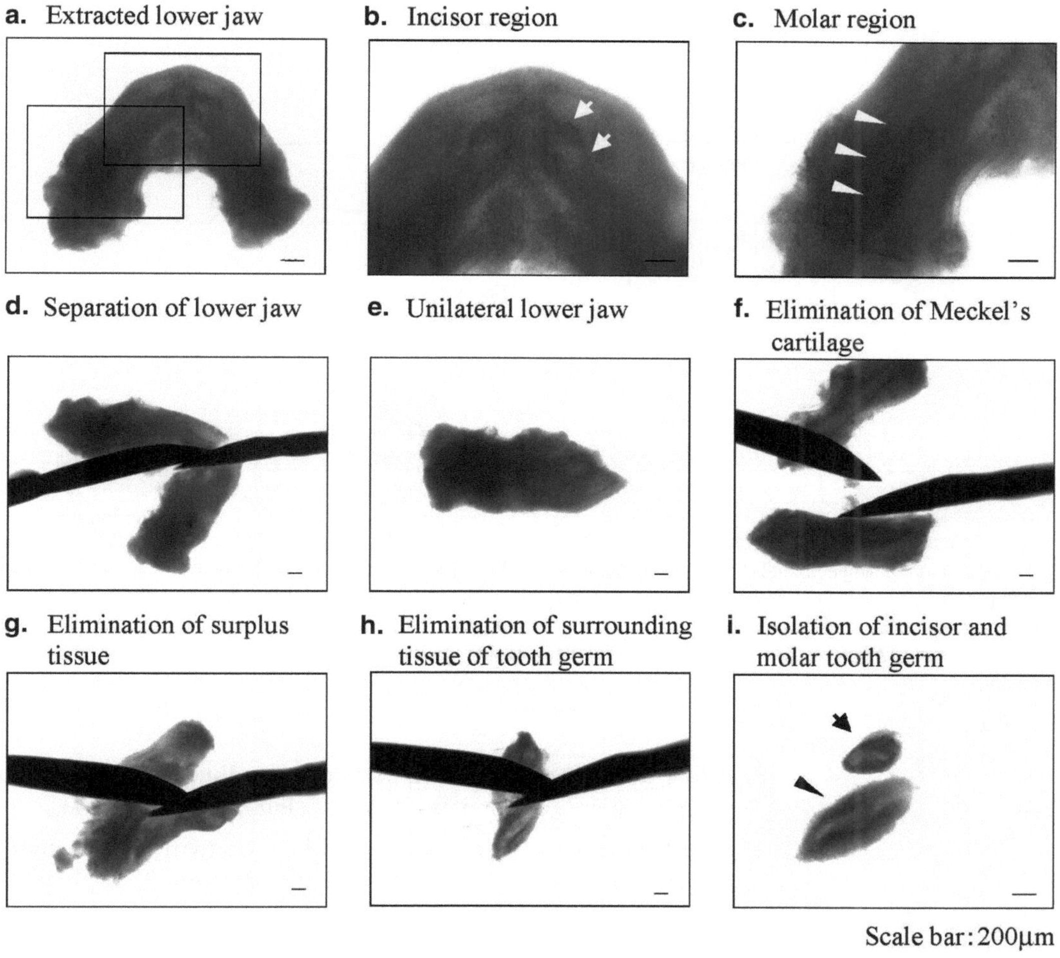

Fig. 1. Dissection of ED14.5 incisor and molar tooth germ. Extraction of the lower jaw from ED 14.5 mouse embryo (**a**). An extended image of the incisor tooth region (**b**) and the molar tooth region (**c**). Separation of left and right lower jaw (**d** and **e**). Removal of Meckel's cartilage (**f**) and excess tissue (**g** and **h**). Isolation of the incisor and molar tooth germ (**i**). The quality of the isolated tooth germ is crucial for the successful development of bioengineered tooth germ. *Arrow*: Incisor tooth germ. *Arrowhead*: Molar tooth germ.

5. Resect all surplus tissue from around an incisor and molar tooth germ (Fig. 1h and i, see Note 20).
6. Keep the isolated tooth germs at 4°C in cold culture medium (see Note 21).

3.2. Separation of Tooth Germ Epithelial and Mesenchymal Tissues

1. Wash the extracted tooth germs twice in PBS(−), add 50 U/mL Dispase solution, and conduct the enzyme reaction at room temperature for 10.5 min (Fig. 2, see Note 22).
2. Stop the enzyme reaction by adding culture medium. Wash the tooth germs twice in the same culture medium (see Note 23).

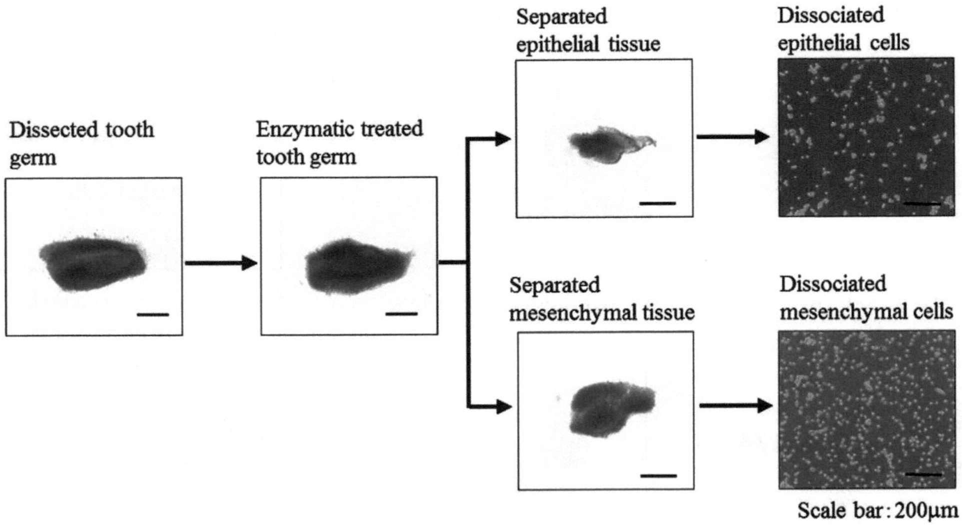

Fig. 2. Preparation of single epithelial and mesenchymal cells. Dissected tooth germs (*left*) treated with 50 U/mL Dispase solution (*second photograph from the left*). Separation of epithelial tissue (*upper, third photograph from the left*) and mesenchymal tissue (*lower, third photograph from the left*) from tooth germ. Preparation of single epithelial cells (*upper, fourth photograph from the left*) and single mesenchymal cells (*lower, fourth photograph from the left*) after enzymatic treatment.

3. Add 1 μL DNase into 2 mL of culture medium and incubate at room temperature for a few seconds (Fig. 2, see Note 24).
4. Separate the tooth germ epithelial and mesenchymal tissues under a microscope using 25-g needles (Fig. 2, see Note 25).
5. Keep each tissue at 4°C in cold culture medium.

3.3. Enzymatic Separation of Single Cells from Tooth Germ Epithelial Tissue

1. Collect the epithelial tissues in a 15-mL tube and centrifuge at 590 × *g* for 3 min. Discard the medium and wash the cell pellet twice in PBS(−).
2. Aspirate to remove the PBS(−) and add 2 mL of enzyme reactive Reagent A. Incubate the epithelial tissues for 15 min in a 37°C water bath. Repeat this procedure twice (see Note 26).
3. Centrifuge the epithelial tissues at 590 × *g* for 3 min and discard all of Regent A. Add 2 mL of enzyme reactive Reagent B and incubate the epithelial tissues for 5 min in a 37°C water bath.
4. Stop the enzyme reaction by adding 6 mL of culture medium. Centrifuge at 590 × *g* for 5 min.
5. Aspirate the supernatant to a residual volume of 80 μL and disperse the cell pellet by tapping (see Note 27).

6. Immediately add 1 mL of culture medium and centrifuge at 590 × *g* for 3 min.
7. Aspirate the supernatant to a residual volume of 200 μL and add 1 μL of DNase solution to the residual volume. Create a single cell suspension by gently pipetting ten times with a micropipette and a P200 tip (see Note 28) and sift the suspension through a cell strainer (Fig. 2).

3.4. Enzymatic Separation of Single Cells from Tooth Germ Mesenchymal Tissue

1. Collect the mesenchymal tissues in a 15-mL tube and centrifuge at 590 × *g* for 3 min. Discard the medium, and wash the cell pellet twice in PBS(−).
2. Aspirate to remove the PBS(−), and add 2 mL of enzyme reactive Reagent C. Incubate the mesenchymal tissues for 10 min in a 37°C water bath.
3. Stop the enzyme reaction by adding 6 mL culture medium and centrifuge at 590 × *g* for 5 min.
4. Aspirate the supernatant to a residual volume of 80 μL and disperse the cell pellet by tapping (see Note 27).
5. Immediately add 1 mL of culture medium and centrifuge at 590 × *g* for 3 min.
6. Aspirate the supernatant to a residual volume of 200 μL and add 1 μL of DNase solution to the residual volume. Create a single cell suspension by gently pipetting ten times with a micropipette and a P200 tip (see Note 28) and sift the suspension through a cell strainer (Fig. 2).

3.5. Reconstitution of the Bioengineered Tooth Germ

1. Prepare siliconized 35-mm petri dishes and 1.5-mL tubes coated with silicon grease (see Note 29).
2. Transfer the epithelial or mesenchymal single cell suspensions isolated from tooth germ into separate siliconized 1.5-mL tubes.
3. Centrifuge at 600 × *g* for 3 min and discard the supernatant using a micropipette and a P1000 or P200 tip.
4. Centrifuge at 600 × *g* for 3 min and discard the residual supernatant on the cell pellets using a micropipette and a gel-loading tip under a microscope (Fig. 3A-a–c, see Note 30).
5. Prepare a droplet of 30 μL collagen gel on a siliconized petri dish (see Note 31).
6. Aspirate a 0.3–0.4-μL volume of the mesenchymal cell pellet using a micropipette and a 0.1–10-μL pipette tip under a microscope (see Note 32). Apply the cell pellet slowly into the collagen drop and make a spherical cell aggregate (Fig. 3B-a–c, see Note 33).

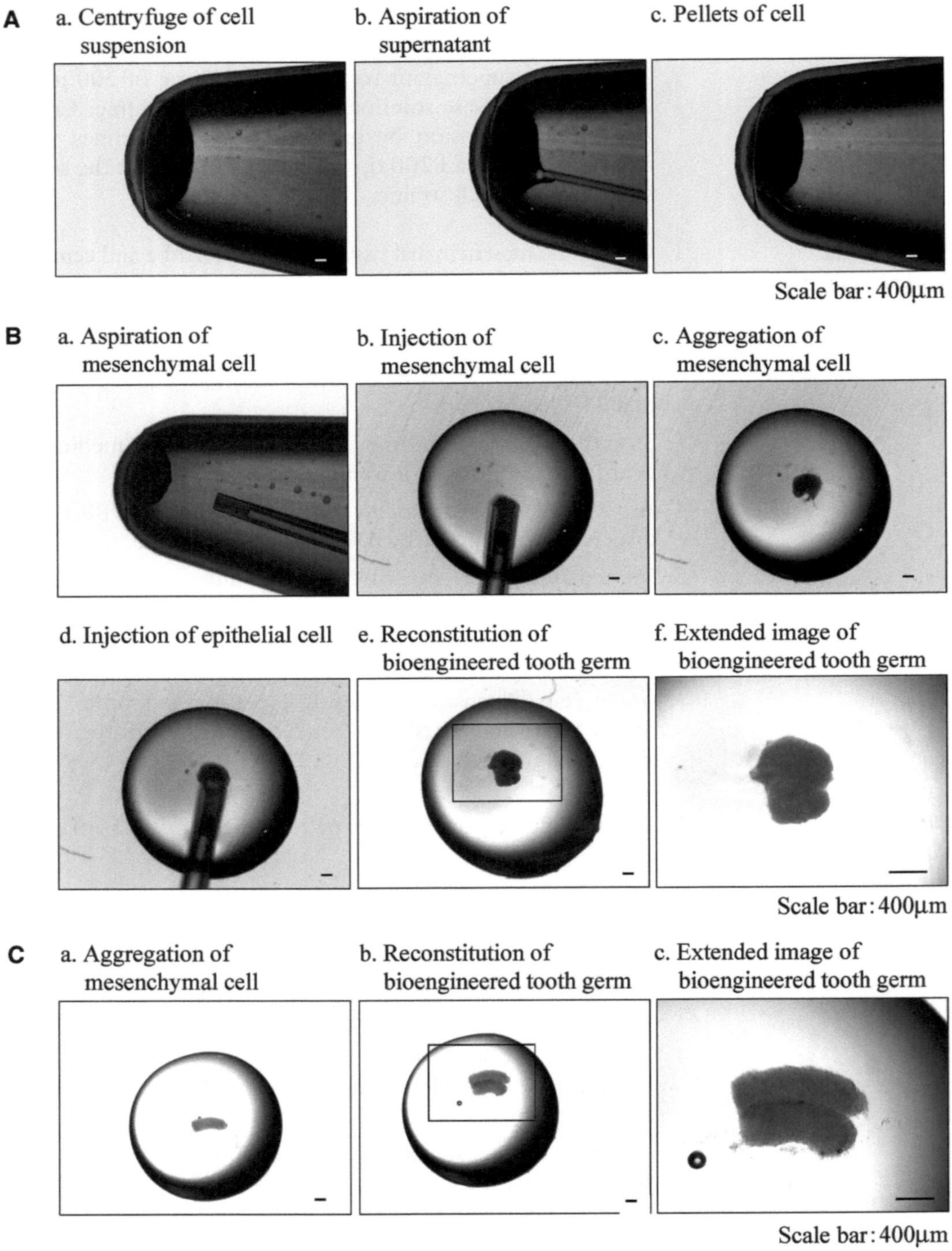

Fig. 3. Reconstitution of bioengineered tooth germ. (**A**) Centrifuge and remove the residual supernatant around the cell pellets (*a–c*). (**B**) Aspirate (*a*) and inject the total volume of mesenchymal cells into the center of the collagen drop (*b* and *c*). Subsequently, aspirate and inject the epithelial cells into the same drop adjacent to the mesenchymal cell aggregate (*d–f*). (**C**) A Hamilton syringe is used during the reconstitution of bioengineered tooth germ to regulate the tooth crown width. Aspirate and inject the preferred volume of mesenchymal cells into the center of the collagen drop (*a*). Subsequently, aspirate and inject the epithelial cells into the same drop adjacent to the mesenchymal cell aggregate (*b* and *c*).

7. Similarly, apply a 0.2–0.3-μL volume of the epithelial cell pellet into the same collagen drop and make contact with the mesenchymal cell aggregate (Fig. 3B-d–f, see Notes 33–35).

3.6. In Vitro Organ Culture of the Bioengineered Tooth Germ

1. Incubate the petri dish holding the collagen gel drop for 15 min at 37°C to coagulate the collagen gel (Fig. 4A-a, see Note 36). Set the cell culture insert into a 12-well plate filled with culture medium (350 μL/well).
2. Pick up the collagen gel drop with tweezers and transfer the drop onto the cell culture insert (Fig. 4A-b and c, see Note 37).
3. Replace the culture medium supplemented with 10% FBS 100 μg/mL ascorbic acid, and 2 mM L-glutamine. Remove the medium and replace with fresh medium every other day (see Note 38).
4. After culture for 14 days, multiple bioengineered teeth will have developed in the collagen gel. Each bioengineered tooth has the correct tooth structure (Fig. 4B).

3.7. Transplantation of a Bioengineered Tooth Germ into the Subrenal Capsule

1. Anesthetize the mouse to be implanted (see Note 39).
2. Place the mouse on its stomach and immobilize the hands and feet.
3. Shave the dorsal hair with a razor (see Note 40).
4. Dissect the dorsal skin in approximately 2-cm in length and exfoliate between the skin and the fascia (Fig. 5A-a).
5. Dissect the fascia and draw out the kidney (Fig. 5A-b).
6. Dissect approximately 2–3 mm of the subrenal capsule outer membrane (Fig. 5A-c, see Note 41).
7. Exfoliate between the kidney parenchyma and the outer membrane with thin tweezers and insert the bioengineered tooth germ sample into the interspace (Fig. 5A-d).
8. Gently return the kidney to the peritoneal cavity. Suture the fascia and skin (Fig. 5A-e, f).
9. Fourteen days after transplantation, multiple bioengineered teeth with the correct tooth structure will have developed in the subrenal capsule (Fig. 5B, C).

3.8. Transplantation of a Bioengineered Tooth Germ into the Transplant Hole

1. Anesthetize the mouse to be implanted (see Note 39).
2. Place the mouse on its back, immobilize the body on the dissecting table, and tie a string to an incisor tooth on the upper and lower mandible.
3. Secure the string in the cheek on both sides to preserve the visual field.

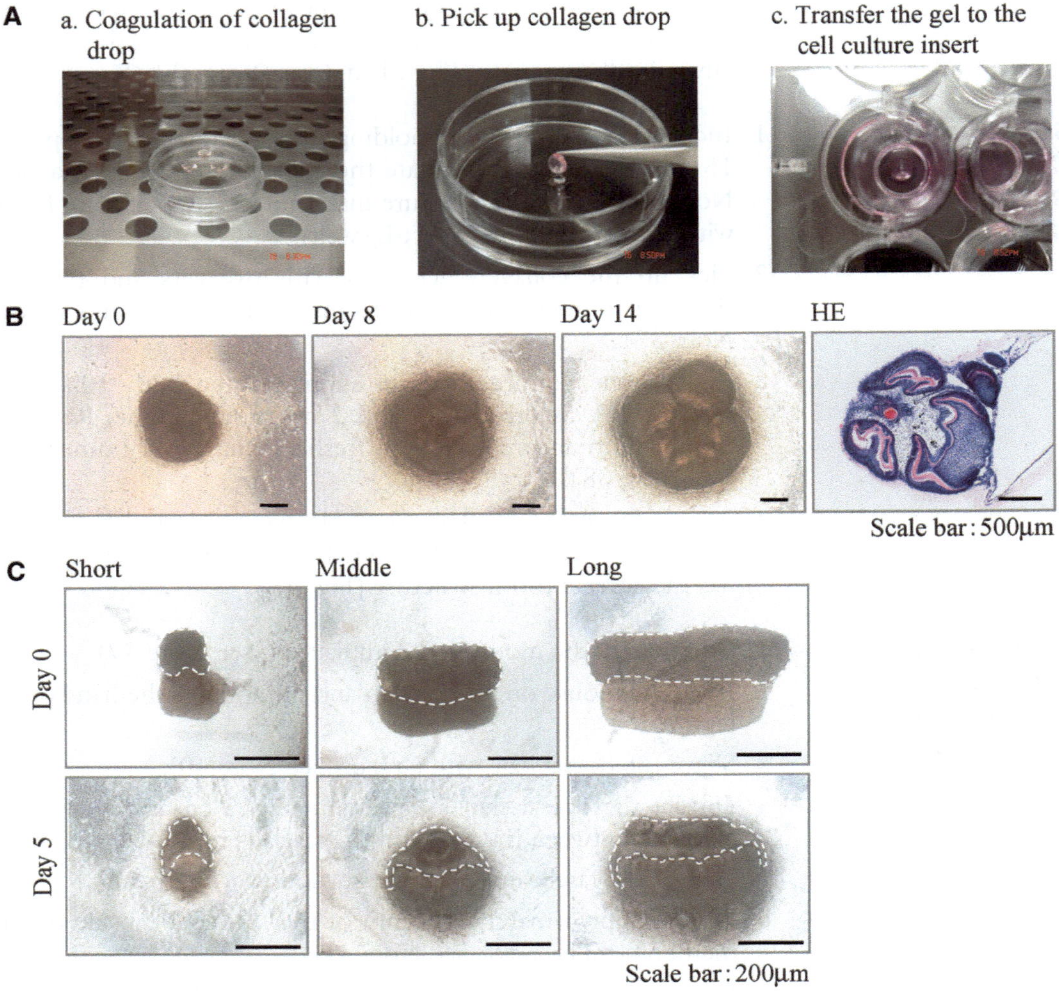

Fig. 4. In vitro organ culture of bioengineered tooth germ. (**A**) Coagulate the collagen gel drop at 37°C (*a*). Pick up the collagen drop (*b*) and transfer onto the cell culture insert (*c*). (**B**) Typical images of bioengineered tooth germ in in vitro organ culture. Phase-contrast image (*first, second, and third photographs from the left*) and Hematoxylin and Eosin staining (*fourth photograph from the left*). (**C**) Phase-contrast images of three-dimensional contact area groups (*left*, short; *center*, middle; *right*, long) of the bioengineered tooth germs after 0 and 5 days of organ culture. All of the bioengineered tooth germs reach the early bell developmental stage following 5 days in organ culture.

4. Cut the periodontal ligament of the upper first molar with a needle (see Note 42).
5. Extract the upper first molar using forceps (Fig. 6B-a–c, see Note 43).
6. Allow 3 weeks after tooth extraction for repair of the oral gingiva and alveolar bone (Fig. 6B-d). Under anesthesia, place the mouse on its back, immobilize the body on the dissecting table, and tie a string to an incisor tooth on the upper and lower mandible.

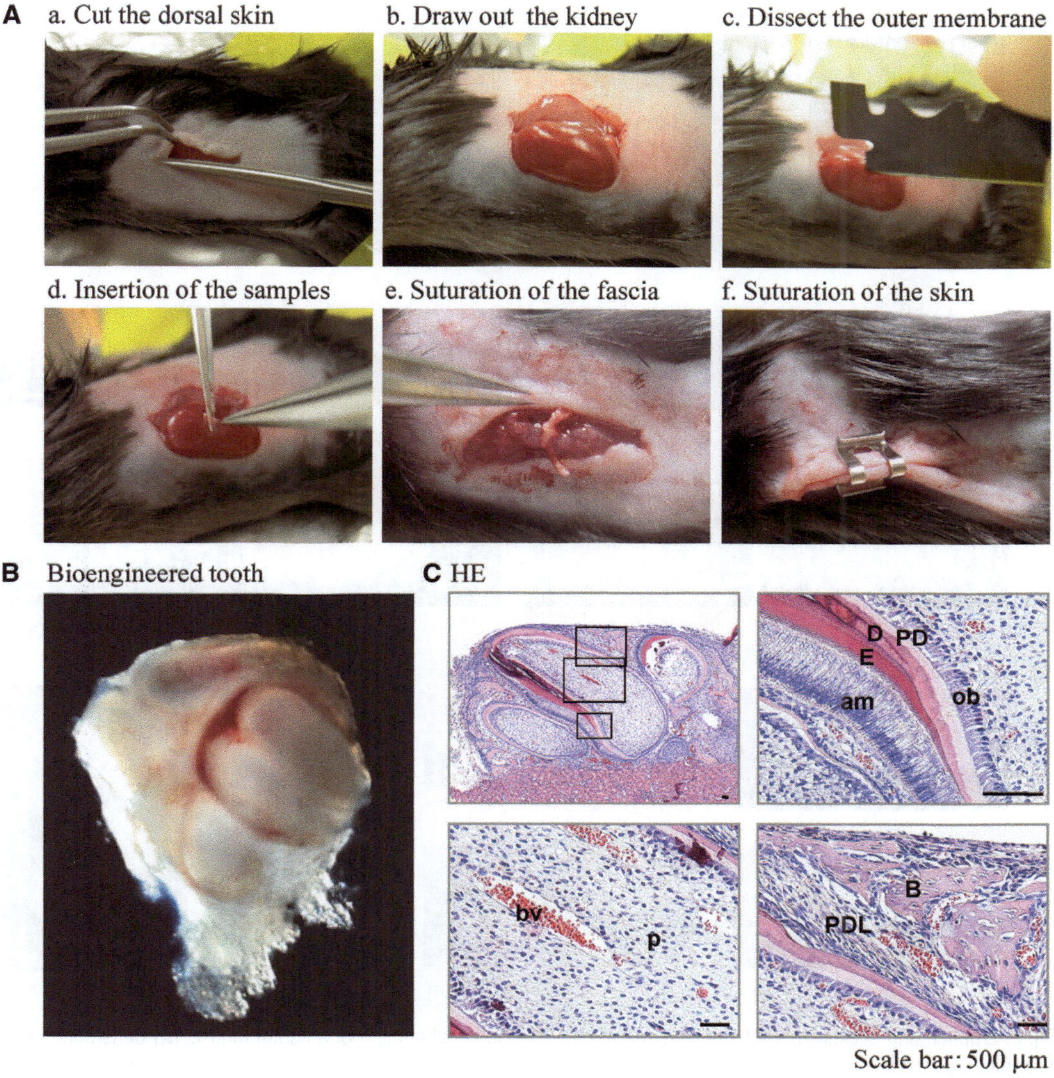

Fig. 5. Development of a bioengineered tooth using subrenal capsule transplantation. (**A**) Make approximately a 2-cm incision on the dorsal skin (***a***) and exfoliate between the skin and the fascia. Draw out the kidney (***b***) and dissect approximately 2–3 mm of the subrenal capsule outer membrane (***c***). Exfoliate between the kidney parenchyma and the outer membrane with thin tweezers and insert the sample into the interspace (***d***). Return the kidney into the peritoneal cavity and then suture the fascia and the skin (***e*** and ***f***). (**B**) Photographs of a bioengineered tooth at 14 days post transplantation into the subrenal capsule. (**C**) Hematoxylin and Eosin staining of a day-14 bioengineered tooth developed in the subrenal capsule. The bioengineered tooth has the same tooth tissue structures as a natural tooth. Ameloblasts (am), odontoblasts (od), pulp cells (p), blood vessels (bv), enamel (E), dentin (D), pre-dentin (PD), periodontal ligaments (PDLs), and alveolar bone (B).

7. Using the string, secure the bilateral buccal mucosa to preserve the visual field.
8. Using a scalpel, incise the oral gingiva at the extraction site and expose the alveolar bone surface (Fig. 6B-e, see Note 44).

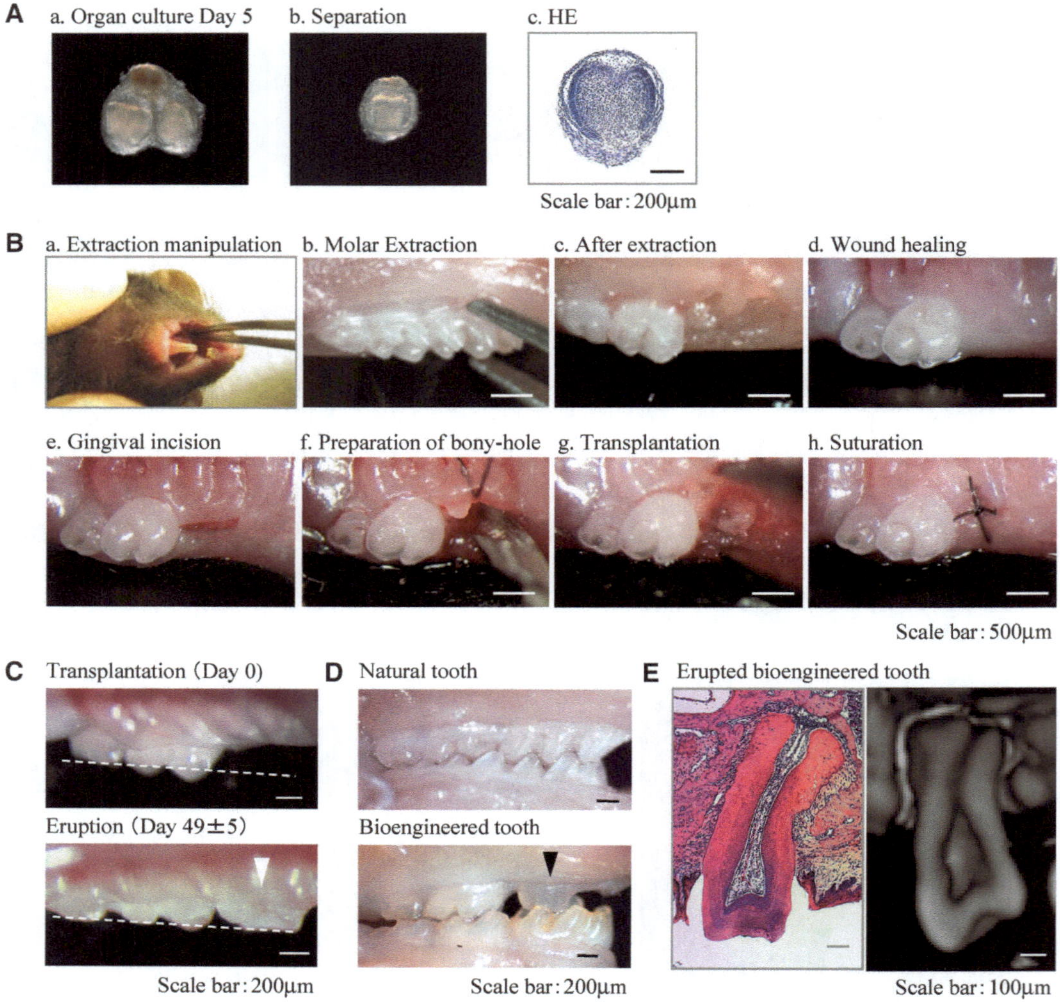

Fig. 6. Eruption and occlusion of a bioengineered tooth in the oral environment of an adult mouse. (**A**) Generation of two bioengineered tooth germs using small-size reconstitution of bioengineered tooth germ (*left*), and by single germ separation using needle manipulations (*center*). The image depicts Hematoxylin and Eosin staining of a single bioengineered tooth germ (*right*). (**B**) Extract the upper first molars under anesthesia (*upper, first, second, and third photographs from the left*). Mice were maintained for 3 weeks to allow for natural repair of the tooth cavity and oral gingiva (*upper, fourth photograph from the left*). Following repair, approximately a 1.50mm-long incision was made through the oral mucosa at the extraction site (*lower, first photograph from the left*). Create a transplant hole of approximately 0.5–10 mm in diameter in the exposed alveolar bone surface (*lower, first photograph from the left*). Transplant a bioengineered tooth germ into the bony hole (*lower, second photograph from the left*). Suture the oral gingiva using 8-0 nylon (*lower, fourth photograph from the left*). (**C**) Photographs of the oral cavity showing a bioengineered tooth during the processes of eruption. *Triangle*: Bioengineered tooth. (**D**) Occlusal states of a natural tooth (*upper*) and a bioengineered tooth (*lower*) following tooth eruption. *Filled triangle*: Bioengineered tooth. (**E**) Hematoxylin and Eosin staining (*left*) and micro CT (*right*) of a bioengineered tooth developed in the mouse jaw bone at day 83 post transplantation. The bioengineered tooth has the same tooth tissue structures as a natural tooth.

9. Create a transplant hole using a φ 0.4-mm drill and expand the transplant hole using a φ 0.8-mm drill (Fig. 6B-f, see Note 45).
10. Transplant a bioengineered tooth germ with the tooth crown side facing upward into the transplant hole (Fig. 6A-g, see Notes 7 and 46).
11. Suture the oral gingiva using 8-0 nylon (Fig. 6A-h, see Note 47).
12. Approximately 45 days after transplantation, the bioengineered tooth will reach the occlusal plane and erupt from the oral mucosa (Fig. 6C). The bioengineered tooth will achieve a central occlusion with the natural lower teeth and exhibit correct tooth tissue structures similar to the mouse's natural teeth (Fig. 6D, E).

4. Notes

1. PBS(−) is used to wash the tissues and cells. Therefore, it should be an isotonic solution that does not cause cell injury.
2. Choose the basal medium after performing the examination experiment (e.g., Dulbecco's modified Eagle's medium). HEPES is added to keep the pH constant, but HEPES is not added into the culture medium for organ culture to avoid any cytotoxicity during extended culture.
3. Dispase dissolves the basal membrane between epithelial and the mesenchymal tissue of the tooth germ. Even when using the same reagents described in this protocol, we recommend evaluating the conditions of temperature and reaction time, since enzymatic activity is easily decreased at 4°C.
4. Collagenase is an enzyme used to digest collagen. Intercellular collagen molecules are resolved by collagenase and tissues are separated into single cells. Even when using the same reagents described in this protocol, we recommend evaluating the conditions of temperature and reaction time, since enzymatic activity is easily decreased at 4°C.
5. Cells can be partially damaged by enzymatic treatment and release their DNA. The DNA released by enzymatic treatment can cause cells to aggregate, making subsequent manipulations difficult. DNase can prevent this cell aggregation by digesting the DNA.
6. Collagen gel can be easily made using Cellmatrix Type I-A (Nitta Gelatin Inc.).
7. When the reconstituted tooth germ is transplanted into a bony hole, it is necessary to consider the direction of tooth eruption.

Use of methylene blue, a biocompatible dye, dissolved in agarose gel to stain the epithelial tissue of the reconstituted tooth germ can facilitate orientation of the bioengineered tooth germ during transplantation.

8. Surgical instruments should be washed and sterilized by autoclaving prior to each use in order to prevent contamination.
9. We recommend the use of minimum-size surgical tools, such as scalpel blades and drills, which can be more easily manipulated in the mouse oral cavity.
10. We recommend the use of a dissecting microscope capable of 6.5–50× magnification with a transmitted beam applied as the light source.
11. A 25-G needle is suitable for most of the manipulations performed during tooth germ extraction.
12. Round-bottom microtubes should be used, since square-bottom microtubes are unsuitable for forming a cell pellet by centrifugation.
13. A 0.1–10-μL pipette tip is suitable for making a highly concentrated cell aggregate in a collagen gel.
14. A membrane should be used with a pore size that is sufficient for liquid components to pass through. We use a cell culture insert/0.4 μm pore size membrane (BD Biosciences).
15. The PIPETMAN® P2 micropipette (Gilson Co. Ltd.) is recommended for facile creation of cell aggregates in the collagen gel.
16. To avoid the risk of bacterial contamination, do not use the same tweezers and scissors for cutting skin and muscle.
17. Mouse embryos can be easily isolated by surgical removal of the uterine and amniotic membranes.
18. Dissecting manipulations of tooth germ should be performed using a pair of 25-G needles. One needle with the cutting surface turned toward the tooth germ is fixed on the sample, while the other needle cuts the tissue by sliding along the cutting surface of the first needle. Care should be taken not to press too firmly on the tissue with the first needle, which can result in a tissue tear.
19. The incisor and molar tooth germ is adjacent to Meckel's cartilage. Careful needle manipulation should be used when separating both bones of jaw from Meckel's cartilage.
20. Correct dissecting tooth germ manipulations can influence the frequency of successful development of a bioengineered tooth germ.
21. Transfer the tooth germ together with the medium. Do not directly transfer a tooth germ in order to avoid injury to the tooth germ.

22. Carefully follow the time and the temperature of enzyme reactions, since long enzyme reactions can injure the tooth germ.
23. Quick and precise washing is required for the Dispase solution.
24. Tissues and cells, which have been enzymatically treated, can easily aggregate due to released DNA, making subsequent manipulations difficult. The addition of DNase digests DNA and can prevent cell aggregation.
25. Needle manipulation should be performed carefully in order to avoid injury to the epithelial and mesenchymal tissue.
26. Tissues can be mixed in order for the enzymatic regents to react equally with several tissues. Care should be taken to avoid tissues sticking to the tube wall.
27. When cell aggregation is not dispersed by tapping, the cell pellet can be manipulated into a single cell suspension by gently pipetting up and down.
28. Micropipette manipulation should be performed gently at a constant speed.
29. Be careful not to apply too much silicon grease and not to apply it to the inside of the 1.5-mL tube cover.
30. Remove as much of the residual supernatant as possible to create high-density cell pellets. If residual supernatant remains, the cell aggregate will not be made into a collagen gel.
31. The cell manipulation, which reconstitutes of a bioengineered tooth germ, should be performed quickly, since the collagen gel solidifies with a change in temperature and with the passage of time.
32. For reconstitution of a bioengineered tooth germ, aspirate only the required amount of cells using a pipette tip.
33. Insert the pipette tip into the collagen gel using a P2 micropipette. A cell aggregate is extruded slowly and the pipette tip must be precisely operated so that cell aggregation becomes spherical. The cell insertion should be stopped precisely when all cells have been extruded from the pipette tip do not extrude air bubbles in to the gel.
34. When bioengineering a tooth germ using both epithelial and/or mesenchymal tooth germ tissues (i.e., reconstitution of tissue and tissue, or tissue and cell), ensure that there is sufficient contact between the tissue and cell aggregate.
35. When you create a bioengineered tooth germ using this cell manipulation method, the number of developed bioengineered tooth germs can be reduced by minimizing the volume for the cell aggregate. Furthermore, the width of the tooth crown can be regulated by controlling the contact area between the epithelial and mesenchymal cell layers (Figs. 3C and 4C). This manipulation is conducted using a Hamilton syringe.

36. Turn the siliconized dish upside down so that the sample does not sink to the bottom of the collagen gel (Fig. 4A-a).
37. Tear off the collagen gel from the bottom aspect of the siliconized dish. The side of the collagen gel should be picked up with tweezers and carefully installed onto the cell culture insert (Fig. 4A-b, c).
38. It is desirable to exchange the entire volume of organ culture medium.
39. Anesthetize the mouse using intraperitoneal injection of 5 mg/mL pentobarbital.
40. Use soapy water for shaving. To avoid the risk of bacterial contamination, shave from the center part of the back to the root of the hind leg in a broad range.
41. When the dissection of the subrenal capsule outer membrane is too large, there is a risk that the outer membrane will ablate from the kidney parenchyma. Care should be taken not to dry out the kidney.
42. A needle is inserted to cut the first molar periodontal ligament of the upper first molar. Be careful not to bore to the maxillary antrum when the needle is inserted deeply.
43. There is a risk of breaking the tooth root when a tooth is pulled up forcibly. The tooth should be extracted by making the tooth vibrate slowly.
44. Incise and exfoliate to create a gingival periosteal flap.
45. The transplant hole should be formed with caution and in consideration of the direction of tooth eruption. The direction of the transplant hole is influenced by the direction of bioengineered tooth eruption.
46. When the bioengineered tooth germ is transplanted into the prepared bony hole, use needle manipulation to separate the single tooth germ (Fig. 6A).
47. Care should be taken not to injure a transplanted tooth germ with the suture needle. Also, suturing should be complete to avoid leakage of the transplanted tooth germ.

References

1. Tucker, A. and Sharpe, P. (2004) The cutting-edge of mammalian development; how the embryo makes teeth. *Nat Rev Genet* 5, 499–508
2. Ikeda, E. and Tsuji, T. (2008) Growing bioengineered teeth from single cells: potential for dental regenerative medicine. *Expert Opin Biol Ther* 8, 735–744
3. Pispa, J. and Thesleff, I. (2003) Mechanisms of ectodermal organogenesis. *Dev Biol* 262, 195–205
4. Fukumoto, S. and Yamada, Y. (2005) Review: extracellular matrix regulates tooth morphogenesis. *Connect Tissue Res* 46, 220–226
5. Saito, M., *et al.* (2009) The KK-Periome database for transcripts of periodontal ligament development. *J Exp Zool B Mol Dev Evol* 312B, 495–502
6. Cai, J., et al. (2007) Patterning the size and number of tooth and its cusps, *Dev. Biol.* 304, 499–507

7. Ishida, K., *et al.* (2011) The regulation of tooth morphogenesis is associated with epithelial cell proliferation and the expression of Sonic hedgehog through epithelial-mesenchymal interactions. *BBRC* 405 455–461
8. Thesleff, I. (2003) Epithelial-mesenchymal signalling regulating tooth morphogenesis. *J Cell Sci* 116, 1647–1648
9. Langer, R.S. and Vacanti, J.P. (1999) Tissue engineering: the challenges ahead. *Sci Am* 280, 86–89
10. Atala, A. (2005) Tissue engineering, stem cells and cloning: current concepts and changing trends. *Expert Opin Biol Ther* 5, 879–892
11. Song, Y *et al.* (2006) Application of lentivirus-mediated RNAi in studying gene function in mammalian tooth development. *Dev Dyn* 235 (5), 1334–44
12. Mantesso, A. and Sharpe, P. (2009) Dental stem cells for tooth regeneration and repair. *Expert Opin Biol Ther* 9, 1143–1154
13. Sharpe, PT., and Young, CS. (2005) Test-tube teeth. *Sci Am* 293:34–41.
14. Brockes, J.P. and Kumar, A. (2005) Appendage regeneration in adult vertebrates and implications for regenerative medicine. *Science* 310, 1919–1923
15. Watt, F.M. and Hogan, B.L. (2000) Out of Eden: stem cells and their niches. *Science* 287, 1427–1430
16. Nakao, K., *et al.* (2007) The development of a bioengineered organ germ method. *Nat Methods* 4, 227–230
17. Ikeda, E., *et al.* (2009) Fully functional bioengineered tooth replacement as an organ replacement therapy. *Proc Natl Acad Sci USA* 106, 13475–13480

Chapter 15

In Vitro Studies on Odontogenic Tumors

Javier Catón, Thimios A. Mitsiadis, and Peter R. Morgan

Abstract

Ameloblastomas are uncommon benign neoplasms of the jaws. They originate from dental epithelial cells, but they are not capable of mineralizing or forming enamel. The study of these tumors is limited to live tissue collected from patients during scheduled surgery. Ameloblastomas grow slowly in vivo and this property is translated to their behavior in vitro. Here, we describe the methods to culture ameloblastomas in organotypic cultures, as well as to isolate stem/progenitor cells from these tumors.

Key words: Ameloblastomas, Odontogenic tumors, Enamel, Organotypic culture, Tumor stem cells, Cell cocultures

1. Introduction

Odontogenic tumors (OTs) present considerable challenges for any investigator willing to use cell and organotypic culture in studies with human tissue as the starting material. These challenges could be summarized as follows:

1. The range and diversity of the tumors
2. The rarity of individual types of odontogenic tumors
3. The frequent, although not exclusive, intra-osseous location of these tumors
4. The diverse tissue composition of the odontogenic tumors
5. The usually slow rate of the odontogenic tumors' growth

If due consideration is paid to these drawbacks, it is possible to employ the tissue in experimental, not simply descriptive, investigations.

Chrissa Kioussi (ed.), *Odontogenesis: Methods and Protocols*, Methods in Molecular Biology, vol. 887,
DOI 10.1007/978-1-61779-860-3_15, © Springer Science+Business Media, LLC 2012

The classification of OTs most used currently is based on that published in 2005 by the World Health Organization (WHO) (1), although unfortunately it introduced several somewhat arbitrary changes in terminologies from those in common use. Broadly, the benign OTs are classified along embryological lines according to whether neoplastic odontogenic epithelium appears to reflect interaction with odontogenic ectomesenchyme or not. One subgroup appears to represent neoplastic growth of tissues derived from the ectomesenchyme itself. Malignant OTs are generally classified descriptively according to their similarity to their benign counterparts.

Benign OTs represent a range of growth disorders from unequivocal neoplasms (e.g., ameloblastomas) to unequivocal hamartomas (compound and complex odontomas) with some entities having an intermediate status (e.g., adenomatoid odontogenic tumor). Parallels with normal tooth development break down with some tumors because these produce unique structures and/or cells not found in the developing teeth (e.g., ghost cells in the calcifying odontogenic cyst, now termed the calcifying cystic odontogenic tumor). These examples illustrate some of the range and diversity exhibited by OTs.

OTs are uncommon tumors. The most common unequivocal OT is the ameloblastoma, a locally aggressive benign neoplasm. Ameloblastomas represent less than 5% of head and neck neoplasms. In the Afro-Caribbean ethnic group, they are more common and in parts of Africa they represent a significant proportion of untreated neoplasms. Controversially, in 2005, the WHO included a cyst, the odontogenic keratocyst, among odontogenic neoplasms, based upon molecular genetic criteria and its propensity for recurrence. This instantly made this cystic lesion the most common odontogenic neoplasm in ethnic Caucasian and Asians, but the status of this entity is still not settled (2). Adenomatoid odontogenic tumors and the odontomes are the next common OTs, after ameloblastomas, both manifesting in a younger population, the second decade. Ameloblastomas and odontogenic keratocysts peak in the fourth or fifth decades. Most other odontogenic tumors are very rare indeed and, unfortunately for the experimental scientist, usually inaccurately diagnosed preoperatively. The commonest malignant OT is the ameloblastic carcinoma, which is more rare than its benign counterpart. A compendium of incidence data for OTs is to be found in Reichart & Philipsen (3).

Most OTs are intra-osseous, a minority of them arising in the gingiva (i.e., peripheral OTs). This makes for difficulties in accessing tumorigenic tissue for culture or banking fresh frozen samples. Band-saw slicing, appropriate for fixed hard tissues, is contraindicated in fresh specimens for reasons of infection control. However, the behavior of the tumors sometimes assists access. In those OTs that expand the jaws, typically large ameloblastomas, the normally dense cortical bone is thinned and can be pried open after slicing

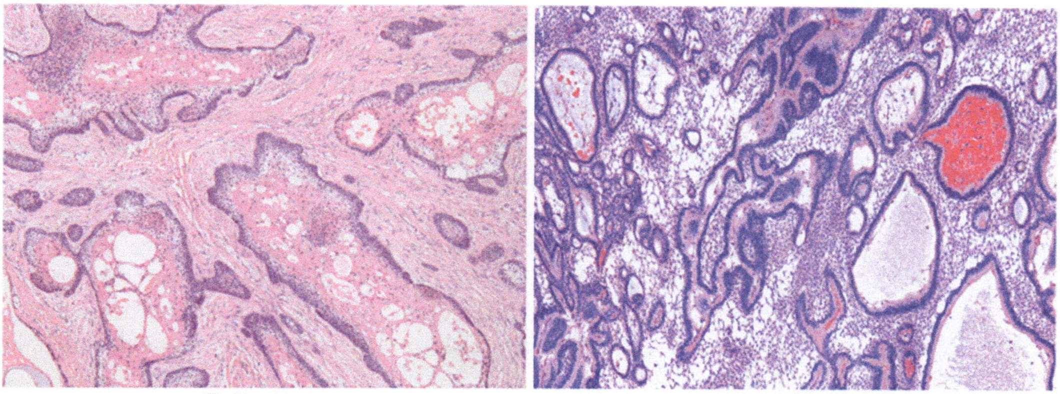

Fig. 1. H&E sections of the two most common solid/multicystic ameloblastomas: Follicular and Plexiform.

with a scalpel to access the soft tumorigenic tissue. Samples taken in this way are less likely to be contaminated with oral microorganisms than those from tumors exposed to the mouth.

A feature of OTs that may pose a problem for diagnosis as well as constituting a disadvantage for cell studies is the extensive cystic change. This is well illustrated in most of the large ameloblastomas. As the neoplasm enlarges, multiple cystic spaces in the epithelial component (i.e., microcysts) and/or the delicate connective tissue (i.e., stromal cysts) expand and coalesce so that opening an expanded cortex reveals a space filled with straw-colored fluid. This is a frequent finding in the two commonest subtypes of ameloblastomas, the solid/multicystic (Fig. 1) and the unicystic variants. Incisional biopsies that include epithelium only from the expanded cyst wall hamper diagnosis, as it is thin and may not show classical features of the tumor and if this is the only material available for cell culture the epithelial cell yield is low.

Other OTs, such as the adenomatoid odontogenic tumor and calcifying odontogenic cyst, may present with a largely cystic expansion of the jaw. The odontogenic keratocyst, assuming we regard it as an OT, has the disadvantage from the perspective of cell culture of not expanding the jaw, or doing so only in juveniles or after a long period of neglect. On the other hand, odontogenic keratocysts are usually treated conservatively nowadays by the delicate detachment of the cyst wall from the inner surface of the jaw (i.e., endosteal), so experimental samples may be obtained direct from the surgeon or pathologist.

Apart from the rare and serendipitous presentation of many OTs, a further problem for cell culture and analytical studies is their heterogeneous nature. Particularly, the rarer OTs contain a mixture of hard and soft tissues. Where the hard tissue is a tooth or several discrete tooth-like elements, these may be removed before explanting or freezing the tissue, but some OTs have dispersed dentine- or enamel-like hard tissues from which odontogenic epithelium, or even mesenchymal stroma, may be impossible to separate. Two of the most common tumors of this type are mature Pindborg tumors (i.e., calcifying epithelial odontogenic tumors) and the cementoblastoma, one category of OT of odontogenic ectomesenchymal origin consisting almost solely of mineralized tissue.

In the following, we describe methods for the culture of ameloblastoma explants taking into consideration the above difficulties for the collection of this tumor. We do not mention any ethical considerations on the understanding that each researcher will follow the protocols of the institution where the research will take place.

2. Materials

Tissue should arrive at the pathology department fresh. The tissue is divided for pathology, tissue bank, and research (see Note 1).

2.1. Cryopreservation

Liquid nitrogen is commercially available (see Note 2), and stored in liquid nitrogen dewars and liquid nitrogen storage containers. Cryo tubes, Cryo 1°C Freezing Container such as "Mr. Frosty," and cryo-protectants such as glycerol or dimethyl sulfoxide (DMSO) for cellular cryopreservation.

2.2. Fixatives

Two different types of fixatives are routinely used: 10% neutral buffered formalin for pathological studies and 4% paraformaldehyde in 0.1 M phosphate buffer (PB) for research.

1. 10% Buffered formalin:

Formaldehyde (37–40%)	100 ml
Distilled water	900 ml
NaH_2PO_4	4.0 g
Na_2HPO_4 (anhydrous)	6.5 g

Mix to dissolve. Store at room temperature.

2. 4% Paraformaldehyde in 0.1 M phosphate buffer.

Paraformaldehyde	40 g
0.1 M Phosphate buffer	1,000 ml

Heat to 60–65°C while stirring. Add a few drops of 1 N NaOH until solution is clear. Continue to stir to dissolve. Cool the solution, filter, and aliquot (see Note 3).

2.3. Buffers

1. 0.2 M Phosphate buffer, pH 7.4.

Na_2HPO_4	21.8 g
NaH_2PO_4	6.4 g

Distilled water up to 1,000 ml.

2. 1 M Tris–Cl (pH 7.4).

Tris base	121.1 g
HCl, concentrated	70 ml

Distilled water up to 1,000 ml.

3. 10× Tris/EDTA (TE).

1 M Tris–Cl	100 ml
0.5 M EDTA, pH 8.0	20 ml

Distilled water up to 1,000 ml (see Note 4).

4. HANKs (commercially available for cell culture).
5. PBS (commercially available for cell culture).

2.4. Tissue Sectioning and Staining

1. SuperFrost Plus glass slides or similar slides to increase adherence.
2. Microtome/cryostat and material related to their use.
3. Wax for embedding (purified paraffin/synthetic resin blend).
4. Xylene/Histoclear (see Note 5).
5. Alcohol gradient, 50, 70, 80, 90, and 100% (see Note 6).
6. Tissue-Tek CRYO-OCT Compound, Sucrose, cryo-embedding molds, and –80°C freezer.
7. Routine H&E staining material.

2.5. Culture Media and Materials

1. Progenitor cell targeted to oral epithelium defined liquid culture medium (CnT24).
2. DMEM supplemented with 1× penicillin/streptomycin antibiotics for washing and tissue transport.
3. DMEM supplemented with 10% fetal calf serum (FCS) and 1× penicillin/streptomycin antibiotics for organotypic cultures.
4. Differentiation medium: BGJb medium supplemented with 10 mg/ml ascorbic acid, 2 mM of sodium ß-glycerophosphate, and 1× penicillin/streptomycin antibiotics.
5. 2.4 U/ml of Dispase.
6. 0.25% Trypsin.
7. Cell and organ culture dishes, metal grids, and support filters.

2.6. Tools for Dissection

1. Glass dish.
2. Forceps, scalpels, blades, and needles.

3. Methods

3.1. Tissue Extraction/ Preservation

Once surgery of the tumor has been scheduled, it is important to follow up with the surgical team. Most hospitals will have a service to bring the tissue to the pathology department (see Note 7), but it is recommended to collect it personally—if possible—to minimize the time from surgery to the laboratory.

3.2. Tissue Selection

A pathologist familiar with ameloblastomas should select an area of the tumor that will be likely to be richer in tumor cells. As mentioned in the introduction, some of these tumors could have very little starting material to work with. The tissue is then divided for pathology, tissue bank (see Note 8), and research. This methods chapter focuses only on the research portion.

3.3. Tissue for Research

The tissue selected for research should be divided:

3.3.1. Fresh Tissue for In Vivo Studies

1. Flash freeze
   This is achieved by submerging the sample in liquid nitrogen or a mixture of dry ice and ethanol. This frozen tissue can be later used to extract nucleic acid for gene expression and genetic studies (qPCR, microarray, etc.).
2. Fixation
   Although some tissue will be fixed using 10% buffered formalin (see Note 9), most tissue for research should be fixed immediately by submerging it in 4% PFA at 4°C overnight. The fixed tissue can be used for genetic studies (generally, we perform in situ hybridization, immunohistochemistry).
   For in situ hybridization, cryosections are normally used.
3. Cryoprotection
   After fixing, the tissue should be rinsed at room temperature in 0.1 M phosphate buffer with 5% sucrose (this process will initiate the cryoprotection of the tissue). Continue with increasing concentrations of sucrose starting at 5% sucrose in phosphate buffer reaching 20% in 5% increments. Proceed for 30 min in each sucrose mixture at room temperature leaving it overnight at 4°C in fresh 20% sucrose/phosphate.
4. Infiltration
   The tissue is then placed into an infiltration mixture (2:1, 20% sucrose phosphate buffer and O.C.T. embedding medium) for 30 min at room temperature before freezing (see Note 10).
5. Embedding and freezing
   Transfer the tissue to an embedding mold and fill the mold with fresh infiltration mixture. Rapidly submerge the mold into Isopentane cooled with liquid nitrogen (see Note 11). After the material is frozen, wrap the block and store at –80°C.

6. Sectioning
   3–5-μm sections are cut at –20°C in a cryostat. To achieve ideal sections, it is critical to have the knife-edge as sharp as possible. Trim the block face to a diamond shape, with the long axis oriented vertically. This orientation helps to make removal of the sections from the knife-edge easier, and will minimize handling damage of the tissue. Use a small camel hairbrush to guide the section off the block face and transfer it to glass slides (see Note 12). Allow the section to dry on the slide at room temperature. Store slides at –80°C until needed.

3.3.2. Fresh Tissue for In Vitro Studies

Fresh tissue can be cultured in an organotypic form to maintain the architectural three dimensions of the tissue or in monolayer cell culture to isolate specific cells.

1. Organotypic cultures
   Tissue is cut into 2–5-mm cubes and placed into Trowell-type culture dishes (4). These cultures can be maintained for approximately 15 days using DMEM medium supplemented with 10% FCS. The tissue can be induced using proteins and/or cell cocultures. The usage of beads for induction has the advantage of showing the effect in the precise site of application. Following are examples of induction beads and cell cocultures.
2. Induction beads
   Affi-gel agarose (75–150-μm diameter) or heparin acrylic beads (100–200 mesh/100–250-μm diameter), depending on the type of proteins, are needed for induction. Ameloblastomas are induced with proteins involved in tooth development. Recombinants are diluted with 0.1% bovine serum albumin (BSA) in PBS, pH 7.4, to concentrations of 100–200 ng/μl. Recombinant BMP2, BMP4 (Fig. 2), and SHH (200 ng/μl) are used to preload affi-gel agarose beads and FGF2, FGF3, and FGF4 (100 ng/μl) are used to preload heparin acrylic beads. These preloaded beads are incubated for 30 min at room temperature and then washed for 5–15 min in culture media before being transferred with a mouth-controlled capillary pipette on top of the explants. As controls, beads loaded with 0.1% BSA in PBS are used. The explants with beads are cultured in serum-free medium for 20 h and processed for in situ hybridization (5, 6), proliferation assays (7), and immunohistochemistry (6, 8) as described in the references (see Note 13).
3. Cell cocultures
   The usage of cell cocultures for induction of ameloblastoma tissues allows exposing the tumors to a collection of factor the cells express in vitro. We use murine embryonic odontogenic cells (pre-ameloblasts or pre-odontoblasts) to determine the effect of each specific cell type on the explant.

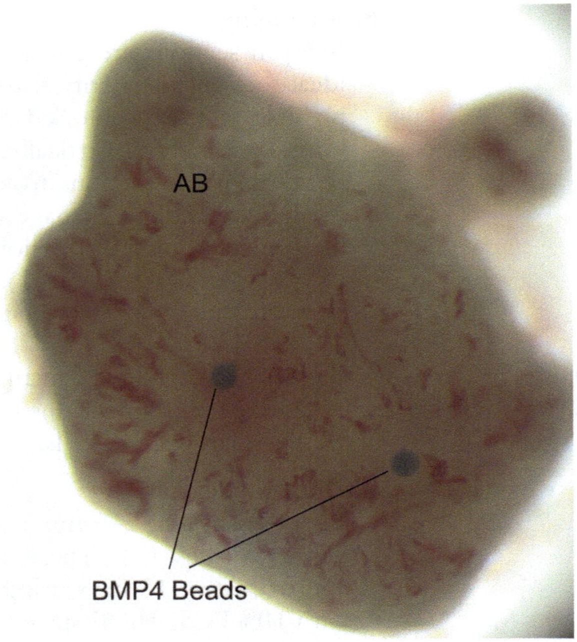

Fig. 2. Preloaded affi-gel agarose beads (BMP4 beads) implanted on top of an ameloblastoma (AB) organotypic culture.

Dissected first mandibular molars from E16.5 mouse (see Note 14) are placed in 2.4 U/ml of Dispase and incubated for 1 h at 4°C. The mesenchymal tissue is mechanically separated from the epithelium using tungsten needles. The two tissues are placed separately in 0.25% trypsin at 37°C for 30 min. Disaggregating is achieved by passing the cells after trypsin treatment through an 18-g needle. The cell suspension is centrifuged and the pellet washed in DMEM with 1× pen/strep. The cells are centrifuged once more and the pellet is placed using a pipette tip in the ameloblastoma explants (Fig. 3). These explants are cultured in differentiation medium for up to 10 days. The cocultured explants are fixed and treated for study as described before.

4. Monolayer cultures

   Cells are isolated from the fresh tissue using two methods:

   Explants shedding allows cells to cast off from small tissue explants into a culture dish. The explants are cut in a similar way as the organotypic cultures. These explants are then placed in 10-cm cell culture dishes with CnT24 media just covering the bottom of the dishes. This allows the explants to have an optimum gas exchange. The explants are kept in this culture conditions for approximately 1 week or when epithelial-like cell colonies are observed attaching to the culture dish. The tissue can also be digested in collagenase IV (freshly prepared

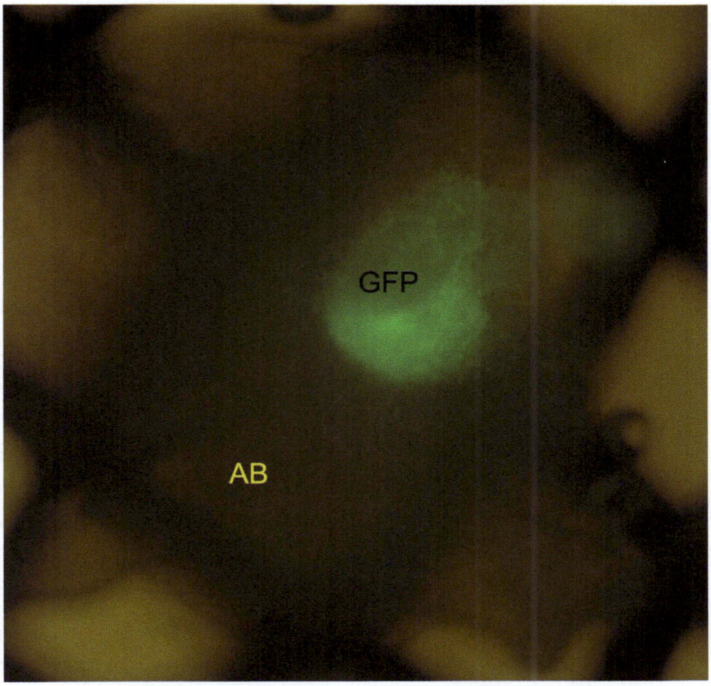

Fig. 3. Odontogenic cells labeled with green fluorescent protein (GFP) placed on top of ameloblastoma (AB) explants in organotypic co-cultures.

500 CDU/ml) at 37°C for 30 min. The cells' disaggregation is achieved by passing them through an 18-g needle. The cell suspension is centrifuged and the pellet washed in DMEM with 1×pen/strep. Centrifuge the cells once more and resuspend the pellet in CnT24 medium for plating in culture dishes (see Note 15).

Ameloblastoma cells can then be characterized for markers common with other epithelial stem cells and other tumor stem cells. These markers can be detected by immunohistochemistry, in situ hybridization, or RT-PCR. For membrane-bound markers, the cells can be shorted using fluorescence activated cell sorting (FACS) (9). This allows separating living cells expressing the marker of interest from the rest.

4. Notes

1. Chemicals are purchased commercially from your choice of provider and solutions should be prepared nuclease free for studies with the need for RNA preservation.

2. Liq N2 should be handled with extreme care and protective clothing, gloves, and eye shield should be worn.
3. 4% PFA aliquots can be stored long term at −20°C. Avoid repeated temperature change cycles and bring to near room temperature before use. It is recommended to use phosphate buffer made with nuclease-free water.
4. We also purchase molecular biology-grade 100× TE for nucleic acid work.
5. Histoclear is less toxic and the results are similar.
6. Dehydration of tissue for nucleic acid work—in situ hybridization—should be done with gradient alcohols made with nuclease-free water.
7. We normally received the tissue from the pathology department. A pathologist selects a portion of the tumor and hands it to the research team in DMEM with pen/strep.
8. Most departments will have a tissue bank for storage of the tissue. This could be useful for in vivo studies.
9. The pathology department that provides the tissue will normally process it in this manner for routine histological analysis. These preparations could be used for morphological study of the tissue.
10. The tissue should sink to the bottom of the container to indicate a correct infiltration.
11. Rapid freezing is recommended, although we have observed that simply placing the mold in dry ice will freeze the sample quickly enough without any adverse effect. It is also possible to use dry ice and ethanol mixture to accelerate the freezing process.
12. If the cryostat has auto-sectioning mode, then one should slow the speed of sectioning (approx. 5 mm/s) to ease the manipulation of the sections as they are coming off the block face. Do not use the anti-roll plate furnished with the cryostat; it compresses the sections and results in poor tissue morphology.
13. In situ hybridization can be done on whole mount or cryosections.
14. We use mice expressing green fluorescent protein (GFP) in order to being able to distinguish the mouse cells from the tumor cells.
15. Epithelial cells are more likely to grow when plated in higher concentrations.

References

1. Barnes, L. et al. (2005) *Pathology and Genetics of Head and Neck Tumours,* IARC Press, Lyon.
2. Li, T. J. (2011) The odontogenic keratocyst: a cyst, or a cystic neoplasm? *J Dent Res 90,* 133–142.
3. Reichart, P. A., and Philipsen, H. P. (2003) [Revision of the 1992 edition of the WHO histological typing of odontogenic tumors. A suggestion], *Mund Kiefer Gesichtschir 7,* 88–93.
4. Trowell, O. A. (1954) A modified technique for organ culture in vitro, *Exp Cell Res 6,* 246–248.
5. Mitsiadis, T. A., Hirsinger, E., Lendahl, U., and Goridis, C. (1998) Delta-notch signaling in odontogenesis: correlation with cytodifferentiation and evidence for feedback regulation, *Dev Biol 204,* 420–431.
6. Mitsiadis, T. A., Salmivirta, M., Muramatsu, T., Muramatsu, H., Rauvala, H., Lehtonen, E., Jalkanen, M., and Thesleff, I. (1995) Expression of the heparin-binding cytokines, midkine (MK) and HB-GAM (pleiotrophin) is associated with epithelial-mesenchymal interactions during fetal development and organogenesis, *Development 121,* 37–51.
7. Mitsiadis, T. A., Muramatsu, T., Muramatsu, H., and Thesleff, I. (1995) Midkine (MK), a heparin-binding growth/differentiation factor, is regulated by retinoic acid and epithelial-mesenchymal interactions in the developing mouse tooth, and affects cell proliferation and morphogenesis, *J Cell Biol 129,* 267–281.
8. Mitsiadis, T. A., Dicou, E., Joffre, A., and Magloire, H. (1992) Immunohistochemical localization of nerve growth factor (NGF) and NGF receptor (NGF-R) in the developing first molar tooth of the rat, *Differentiation 49,* 47–61.
9. Mekada, E., Yamaizumi, M., and Okada, Y. (1978) An attempt to separate mononuclear cells fused with human red blood cell-ghosts from a cell mixture treated with HVJ (Sendai virus) using a fluorescence activated cell sorter (FACS II), *J Histochem Cytochem 26,* 62–67.

Chapter 16

Whole Mount Immunohistochemistry and In Situ Hybridization of Larval and Adult Zebrafish Dental Tissues

Barbara Verstraeten, Ellen Sanders, and Ann Huysseune

Abstract

Tooth development is increasingly being studied in a variety of vertebrate model organisms, each contributing its own perspective to our understanding of dental diversity. In situ hybridization and immunohistochemistry are well-established and frequently used techniques to study the presence of mRNA and protein. Here, we describe a protocol for whole mount immunohistochemistry and in situ hybridization that can be applied to all stages of zebrafish development and dissected bony parts. The description of these protocols is followed by the outline of a quick decalcification method and the procedure for embedding in epoxy resin to obtain serial sections with high histological quality.

Key words: Tooth development, Tooth replacement, Zebrafish, Whole mount immunohistochemistry, Whole mount, In situ hybridization, Epon embedding, Decalcification

1. Introduction

Over the past years, the zebrafish dentition which is restricted to the last pharyngeal arch has been increasingly used to study various aspects of tooth development and replacement, such as the involvement of stem cells (1), role of transcription factors (2), involvement of various signaling pathways (3–5), or distribution of adhesion molecules (6, 7). It has also emerged as a model in regenerative dentistry (8). Stock (2007) provides an excellent overview describing various features of the zebrafish pharyngeal dentition within a comparative context and illustrating how zebrafish can be used to explore the developmental genetic mechanisms of vertebrate dental evolution (9).

As is the case for mammals, studying gene expression and protein distribution constitutes a major tool to advance our knowl-

Chrissa Kioussi (ed.), *Odontogenesis: Methods and Protocols*, Methods in Molecular Biology, vol. 887,
DOI 10.1007/978-1-61779-860-3_16,

edge on how teeth are formed and, specifically in zebrafish, replaced continuously. Gene expression and protein distribution data can validate microarray results and RT-PCR data, and serve as a platform for functional studies. While in mammals gene and protein expression data are routinely collected from sections of paraffin-embedded tissue or on cryosections, the requirements for zebrafish teeth are different. This is largely because of the extreme small size of the first-generation (primary) teeth in zebrafish, which are only approximately 12 μm across at their broadest point (i.e., the thickness of one cryosection, or one to two paraffin sections) and contain less than a dozen odontoblasts in the pulp cavity. With such a small organ size, the level of histological detail that can be obtained quickly becomes critical, and losing one or a few sections rapidly turns the sample useless.

The zebrafish is widely used as a vertebrate model for developmental studies, one of its advantages being its fast embryonic development within the translucent chorion. Thus, not surprisingly, in situ hybridization and immunohistochemistry whole mount protocols have been developed in numerous variations. The advantage of whole mount methods is their speed, as results can be examined immediately following the completion of the protocol using a stereomicroscope. This method is appropriate for early developmental stages (up to 5–6 days post-fertilization), and in the case of teeth it allows to identify the presence or absence of first-generation tooth germs, albeit on a rather coarse level. However, it is largely insufficient to identify the cells or cell layers that express a gene or a protein. As the specimens become larger (larvae of 1–2-weeks old) and considering that the teeth develop close to the central axis of the body, whole mounts do not give any accurate information anymore and sections need to be made. One way of achieving this is by confocal laser scanning microscopy. This approach yields optical sections and stacks, but has the downside that the resolution diminishes if the specimen becomes larger and the tissue of interest is situated deep inside the body, as is the case with the pharyngeal teeth. Moreover, it does not provide the histological detail required to localize transcripts or proteins to a certain cell type or tissue layer. For even older and/or larger fish (juveniles and adults), the whole mount protocol fails to work altogether.

Here, we present a whole mount protocol for immunohistochemistry and in situ hybridization on zebrafish embryos, larvae, and dissected juvenile and adult tissue adapted from the protocol of Dr. P. Raymond (10). This protocol combines the ease of whole mount procedures with high-resolution histology to identify gene expression patterns and protein localization. The essential steps in the protocol include appropriate fixation, decalcification, whole mount in situ hybridization, or immunohistochemistry, followed by embedding in epon (11) and sectioning at any desired thickness.

We have tested the efficiency of this protocol using probes for signaling molecules, transcription factors, and structural proteins, as well as for various antibodies.

The protocols described below are applicable for the study not just of teeth, but also of other organs or organ systems that develop late in zebrafish, such as the bony skeleton (12). They allow us to combine expression data with high-resolution localization of transcripts and proteins.

2. Materials

2.1. Anesthesia

1. 1% MS-222 stock solution: Dissolve 1 g of MS-222 powder in 100 mL distilled water. Store at 4°C.

2.2. Fixation of Zebrafish Embryos, Larvae, and Adults

1. 1×phosphate-buffered saline (PBS): Dissolve two PBS tablets to 1 L distilled water. Store at room temperature (RT) for several months.
2. 4% paraformaldehyde (PFA): Dissolve 4 g paraformaldehyde powder in 100 mL 1× PBS (see Note 1). Preferably, always prepare fresh.

2.3. Depigmentation

1. Depigmentation mix: Add to 8.5 mL distilled water 500 μL 30% H_2O_2 and 1 mL 5% KOH solution. Always prepare fresh.

2.4. Whole Mount Immuno-histochemistry (No RNase-Free Solutions Required)

1. 1× PBST: 0.1% Tween-20 in 1× PBS. Store at RT for several months.
2. Proteinase K (PK) stock solution: Dissolve 1 g proteinase K powder in 100 mL 1× PBS. This is a 1,000× stock solution with 10 mg/mL concentration. Aliquot 0.5 mL in eppendorfs and store at −20°C for maximum 1 year.
3. Blocking solution: 1% Bovine serum albumin (BSA) and 1% dimethylsulfoxide (DMSO) in 1× PBS (see Note 2). Always prepare fresh.
4. StreptABComplex (Dako, Glostrup, Denmark): Add to 1 mL 1× PBS: 5 μL Streptavidin (A) and 5 μL horseradish peroxidase (HRP) (B) (see Note 3). Always prepare fresh.
5. Liquid DAB Substrate pack (Biogenex, Fremont, CA, USA): Add two drops of H_2O_2 solution, four drops of DAB chromogen solution, and 500 μL 10× substrate buffer solution to 4.5 mL distilled water (see Note 4). Always make fresh.

2.5. Whole Mount In Situ Hybridization

All solutions are prepared using RNase-free 1× PBS at room temperature, unless stated otherwise.

Always work with RNase-free tips and eppendorfs.

2.5.1. Prehybridization

1. RNase-free DEPC-water: Add 1 mL DEPC to 1 L distilled water (see Note 5). Leave overnight at RT or 1 h at 37°C, shaking. Autoclave before use to inactivate. Store at RT for several months.
2. RNase-free 1× PBS: Dissolve two PBS tablets to 1 L DEPC-water. Store at RT for several months.
3. 1× PBST: 0.1% Tween-20 in RNase-free 1× PBS. Store at RT for several months.
4. Methanol–xylene mixture: Add equal amounts of 100% methanol (MeOH) and xylene and mix well (see Note 6). Make fresh.
5. Methanol series: Always dilute with RNase-free 1× PBS. Store at RT.
6. Proteinase K stock solution: Dissolve 1 g proteinase K powder in 100 mL 1× PBS. This is a 1,000× stock solution with 10 mg/mL concentration. Aliquot 0.5 mL in eppendorfs and store at −20°C for maximum 1 year.
7. 4% PFA: Dissolve 4 g paraformaldehyde powder in 100 mL 1× PBS (see Note 1). Preferably, make fresh.
8. 0.1 M Triethanolamine (TEA): 6.7 M stock solution. Dilute 67 times by adding 750 μL stock solution to 50 mL DEPC-water. Store at RT for several months.
9. Acetic anhydride/TEA: Add 13 μL acetic anhydride (Sigma, St. Louis, MO, USA) to 10 mL 0.1 M TEA (see Note 7). Always prepare fresh.
10. Hybridization solution (25 mL): 12.5 mL deionized formamide, 6.25 mL 20×SSC, 25 μL heparin (50 mg/mL), 25 μL Tween-20, 25 μL yeast tRNA (50 mg/mL), and 6.2 mL DEPC-water. Store at −20°C for months.

2.5.2. Posthybridization

1. 2×SSC: Starting from 20×SSC, add 10 mL 20×SSC to 90 mL distilled water. Store at RT.
2. 5×MAB buffer: Dissolve 11.6 g maleic acid disodium salt in 100 mL distilled water. Adjust pH to 7.5 with around 600 μL 1 N HCl. Add 53 mL 5 M NaCl and make up to 200 mL with distilled water. Autoclave and store at RT for several months.
3. Maleate blocking solution: 9.8 mL 1×MAB buffer (diluted from 5×MAB buffer with distilled water), 0.2 g Blocking Reagent, 0.2 g BSA powder, 200 μL sheep serum, 10 μL Tween-20 (see Note 8). Always make fresh.

2.5.3. Color Reaction

1. Genius buffer: 1 mL 1 M Tris–HCl (pH 9.5), 500 μL 2 M NaCl, 500 μL 1 M MgCl, 10 μL Tween-20, and 7.9 mL distilled water (see Note 9). Always prepare fresh.
2. Nitro blue tetrazolium (NBT)/5-bromo-4-chloro-3-indolyl phosphate (BCIP): Dissolve 50 mg NBT powder in 700 μL *N*,*N*-dimethylformamide (DMF) and 300 μL distilled water (store at −20°C). Dissolve 50 mg BCIP powder into 1 mL DMF anhydrous (store at −20°C). Add 45 μL NBT and 70 μL BCIP to 10 mL Genius buffer. Keep protected from light.

2.6. Epon Embedding

1. 25% Decalc solution: Add 750 mL of distilled water to 250 mL Decalc. Store at RT.
2. Epon A: 6.2 mL epoxy embedding medium and 10 mL epoxy embedding medium, hardener DDSA (see Note 10).
3. Epon B: 10 mL epoxy embedding medium and 8.9 mL epoxy embedding medium, hardener MNA (see Note 10).
4. Soft epon: 6 mL Epon A, 4 mL Epon B, and 190 μL epoxy embedding medium, accelerator DMP30 (see Note 11). Always make fresh.

3. Methods

All steps are performed at room temperature, unless stated otherwise. Embryos and tissues are kept in eppendorfs throughout the protocol, provided the eppendorfs are gently swirled after each change of solution.

3.1. Fixation of Zebrafish Embryos, Larvae, and Dissected Pharyngeal Jaws of Juveniles and Adults

1. Euthanize the embryos or larvae by adding 1 mL 1%MS-222 in 5 mL aquarium water or 10 mL 1% MS-222 in 50 mL aquarium water for adult zebrafish.
2. If necessary, dechorionate the zebrafish embryos (see Note 12).
3. Dissect the pharyngeal jaws from juvenile or adult zebrafish using forceps and micro-scissors.
4. Transfer the embryos, larvae, or dissected tissue into an eppendorf containing 4% PFA and fix overnight at 4°C.

3.2. Depigmentation (Required for All Developmental Stages, Except for Dissected Pharyngeal Jaws)

1. Remove the 4% PFA and rinse the embryos twice with 1× PBS.
2. Add 1 mL of depigmentation mix to the embryos or larvae and expose to a spotlight for 10–15 min (see Note 13).
3. Once pigments are lost, wash the embryos two times for 5 min with 1× PBS.

4. Finally, transfer them through an ascending MeOH series diluted in 1× PBS (25, 50, 75, and 100%, each step for 5 min). They can now be stored at −20°C for months.

3.3. Whole Mount Immunohistochemistry (Fig. 1a, c)

Day 1

1. Rehydrate the embryos or tissue by transferring them through a descending MeOH series diluted in 1× PBS (75, 50, and 25%, each step for 5 min).
2. Wash two times for 5 min with 1× PBST.
3. Permeabilize the embryos, larvae, or adult tissue with proteinase K in an age-dependent concentration according to Table 1. This is the most critical step of the protocol and has to be performed at the concentration and time as described. A concentration or incubation time that is too low will compromise the penetration of the antibody. A concentration or incubation time that is too high will lead to the destruction of the tissue and the deterioration of the embryos (see Note 14).

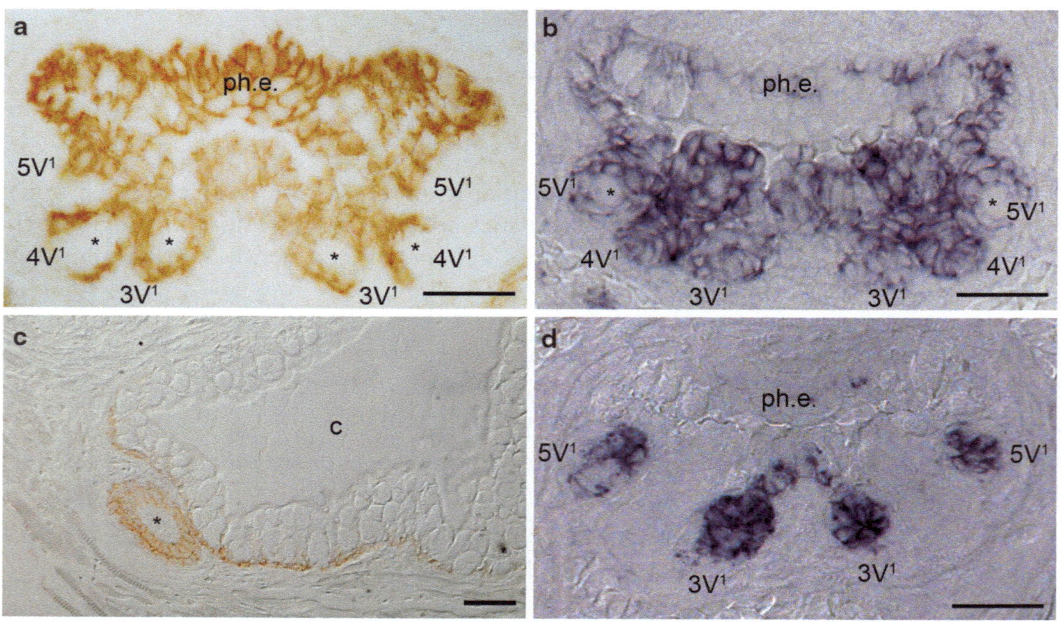

Fig. 1. Whole mount immunohistochemistry and whole mount in situ hybridization results obtained using the protocols described in this chapter. Four microns cross sections of the pharyngeal region of larval zebrafish (**a**, **b**, **d**) and dissected pharyngeal jaws of adult zebrafish (**c**) after whole mount immunostaining (**a**, **c**) or in situ hybridization (**b**, **d**). (**a**) E-cadherin distribution in an 80-h post-fertilization (hpf)-old zebrafish larva showing E-cadherin present in the enamel organ of each first-generation tooth ($3V^1$, $4V^1$, $5V^1$; tooth coding as described in ref. 13); (**b**) whole mount in situ hybridization of an 80-hpf zebrafish larva showing E-cadherin mRNA expression; (**c**) E-cadherin distribution in morphogenesis stage of a later generation tooth obtained from immunohistochemistry on a dissected adult pharyngeal jaw; (**d**) sonic hedgehog (*shh*) expression in first-generation teeth at 72 hpf; note the absence of expression in late cytodifferentiation stage of $4V^1$. *Asterisk* dental papilla, *c* crypt, *ph.e.* pharyngeal epithelium. Scale bars = 25 μm.

Table 1
Overview of concentration and duration of proteinase K (PK) treatment in relation to larval age (in hours or days post-fertilization, hpf and dpf, respectively) and dissected pharyngeal jaws of juvenile and adult zebrafish

Age	PK concentration Diluted from 1,000× stock solution	Time
36–48 hpf	1 × PK	40 min
2.5 dpf	1.5 × PK	40 min
3 dpf	2 × PK	40 min
4 dpf	3 × PK	40 min
5 dpf	4 × PK	40 min
6 dpf	5 × PK	40 min
7 dpf	6 × PK	40 min
Pharyngeal jaw	7 × PK	40 min

4. Remove the proteinase K by washing two times for 5 min with 1× PBST.
5. Block aspecific binding by incubating in blocking solution for 2 h (see Note 15).
6. Dilute the primary antibody in the blocking solution (see Note 16).
7. Incubate the embryos, larvae, or tissue with primary antibody overnight at 4°C. Meanwhile, the corresponding negative control remains in blocking solution also overnight at 4°C.

Day 2

8. Wash four times for 30 min with blocking solution (see Note 17).
9. Dilute the secondary antibody in blocking solution (see Note 16).
10. Incubate all stages (including negative controls) with the secondary antibody overnight at 4°C.

Day 3

11. Wash three times for 30 min with blocking solution (see Note 17).
12. Make StreptABComplex and let it rest for 30 min at RT before use.
13. Wash one more time for 30 min with blocking solution (see Note 17).
14. Incubate in StreptABComplex for 45 min, never more than 1 h.
15. Wash three times for 30 min with blocking solution (see Note 17).

16. Wash one time for 30 min with 1× PBST.
17. Incubate in freshly prepared liquid DAB solution (see Note 4). For a given primary antibody, always start with determining the optimal incubation time for one developmental stage by checking at regular intervals using a stereomicroscope. Use the same incubation time for all stages in order to allow comparisons.
18. To stop the color development and rinse away the DAB, wash three times for 5 min with 1× PBS. Hereafter, the samples are ready to be embedded in epon (see Subheading 3.5).

3.4. Whole Mount In Situ Hybridization (Fig. 1b, d)

Day 1: Prehybridization

1. Rinse in 100% MeOH for 5 min.
2. Rinse in 1:1 100% MeOH and xylene for 5 min.
3. Rinse in 100% xylene for 30 min.
4. Rinse in 100% MeOH for 30 min.
5. Rehydrate tissue in a descending methanol series in 1× PBST (90, 70, and 50%; each step for 5 min).
6. Rinse in 1× PBST two times, 15 min each.
7. Make an appropriate concentration of proteinase K in PBST starting from the 1,000× stock solution. This treatment is age dependent and should be performed according to Table 1 (see Note 14).
8. Rinse embryos or tissue for 10–15 s in 1× PBST.
9. Re-fixation: Add 4% PFA in PBS for 20 min.
10. Wash with 1× PBST twice for 5 min.
11. Rinse in 0.1 M TEA for 3 min.
12. Incubate tissue in acetic anhydride/TEA solution for 10 min.
13. Rinse for 30 min in 1× PBST.
14. Prehybridize in 500 μl of prewarmed hybridization solution for 1–2 h at 65°C.
15. Prepare DIG-labeled probes (both sense and antisense probes): Dilute to known working concentration in hybridization solution (see Note 18).
16. Heat the diluted probe at 80°C for 10 min and put on ice to cool down quickly.
17. Remove prehybridization solution from eppendorfs and add hybridization solution containing diluted probe.
18. Hybridize overnight at 65°C.

Day 2: Posthybridization

19. Prewarm the following solutions: 1:1 formamide/2×SSC at 65°C and 10 mL 2×SSC at 37°C.
20. Rinse briefly in 2×SSC at RT.

21. Wash for 1 h in 1:1 formamide/2×SSC at 65°C.
22. Wash two times for 15 min each in 2×SSC at 37°C (see Note 19).
23. Wash in 1× PBST for 15 min at RT.
24. Block with maleate blocking solution for 2 h at RT.
25. Dilute α-DIG-alkaline phosphatase-coupled antibody in maleate blocking solution (see Note 16), and incubate overnight at 4°C.

Day 3: Color reaction

26. Start with a maleate blocking solution wash, three times for 10 min each.
27. Incubate two times for 5 min each in Genius buffer.
28. Transfer embryos or tissue to a 24- or 96-well plate.
29. Prepare NBT/BCIP solution and add 200 μL per well.
    Incubate at RT in the dark (wrap the well plate in aluminum foil). Monitor reaction frequently using a stereomicroscope.
30. Once the signal is sufficiently strong, wash in 1× PBST two times for 15 min at RT to stop the reaction.
31. Fix in 4% PFA for 30 min at RT or overnight at 4°C.
32. Rinse for 5 min in PBS to remove the PFA. Hereafter, the samples are ready to be embedded in epon (see Subheading 3.5).

3.5. Epon Embedding

Once the whole mount immunostaining or in situ hybridization is finished, the embryos, larvae, or tissues can be embedded in plastic to make histological sections. For zebrafish, up to 7 days post-fertilization (dpf), a decalcification step is not needed and the embedding protocol can be started at the dehydration step (step 3). Older specimens and dissected pharyngeal jaws need to be decalcified before embedding to enable sectioning.

1. Add 1 mL of 25% Decalc to the tissue in an eppendorf and leave for 1 week at RT, changing the Decalc solution on a daily basis.
2. Wash larvae in several steps of 1× PBS; large zebrafish or dissected pharyngeal jaws can also be washed overnight in running tap water in a plastic biopsy cassette.
3. 30% EtOH for 15 min.
4. Remove half of the volume and add an equal amount of 50% EtOH so that you obtain 1:1 30%/50% EtOH. Leave the sample in this solution for 15 min.
5. 50% EtOH for 20 min.
6. 50%/70% EtOH for 20 min.
7. 70% EtOH for 30 min.
8. 70%/95% EtOH for 30 min.

9. 95% EtOH for 30 min.
10. 95%/100% EtOH for 30 min.
11. 100% EtOH two times for 1 h.

To be able to observe the colored precipitate, sections are preferably made at 4 μm. Therefore, the so-called soft epon is used to embed the specimens.

1. Prepare soft epon (see Notes 10 and 11).
2. Add equal amounts of soft epon and 100% EtOH to a glass jar.
3. Gently stir the mixture (to avoid air bubbles) until epon and ethanol are mixed well.
4. Add the dehydrated samples to the epon–ethanol mixture and close the jar from the air for 48 h; keep at RT.
5. After 48 h, open the jar under the fume hood overnight so that the ethanol can evaporate.
6. Next day, prepare fresh soft epon.
7. Transfer the embryos, larvae, or tissue into a rubber mold containing freshly made soft epon.
8. Place the mold in a vacuum oven at 37°C under a partial vacuum of 400 mmHg for 4 h. In this way, trapped air bubbles can escape from the epon.
9. For the final embedding, transfer the larvae to the definite mold with soft epon and position them as you like. Put the mold in the oven at 60°C in order to polymerize the epoxy resin. It will take 2 days to fully harden.
10. Take the blocks from the mold. They are ready to be sectioned with a glass or a diamond knife on a standard microtome.

4. Notes

1. To dissolve the powder, gently stir with a magnetic stirrer while heating up to 70°C. Never boil the solution. It is advisable to use finest grain powder as it will dissolve without adding NaOH. If the powder does not dissolve well, add one to two drops of 1 N NaOH to change the pH and enhance the dissolution. To fix specimens for in situ hybridization, always dissolve PFA powder in RNase-free 1× PBS (see Subheading 2.5).
2. Make this blocking solution in a falcon approximately 15 min before use. By using a falcon, it is possible to shake and vortex the solution in order to dissolve the BSA flakes quicker. Prepare this blocking solution fresh on the first day of the protocol. It

will be used to dilute the primary and secondary antibodies and can be stored at 4°C during the time span of the protocol.

3. Prepare this mixture 30 min before use to allow appropriate complex formation. This kit combines streptavidin and HRP. Streptavidin has a high affinity for the biotin-coupled secondary antibody. At the sites where the primary antibody is bound to the targeted protein, HRP will convert the chromogen DAB into a brown precipitate.
4. Always wear protective clothing while using DAB: it is very carcinogenic. If spilled, clean with bleach to deactivate. After use, dispose into the appropriate disposal bin.
5. DEPC is carcinogenic. Always manipulate under the fume hood. Avoid contact with skin by wearing protective clothing. Once dissolved in water and autoclaved, these precautions are no longer necessary.
6. Be aware that xylene can dissolve certain plastics. Therefore, always use eppendorfs or falcons which are xylene resistant. You can always use glassware, but make sure it is RNase-free. Always make this mixture fresh.
7. Manipulate under the fume hood as acetic anhydride is irritant and the vapor is harmful.
8. Mix the Blocking Reagent with 1×MAB and stir with a magnetic stirrer on a hot plate (60°C) until the powder has dissolved. Let it cool down at RT and add the Tween-20, BSA powder, and sheep serum.
9. Genius buffer is always made fresh and diluted with distilled water. When diluted with another buffer, it will precipitate.
10. Mix the components well by stirring gently with a magnetic stirrer. Avoid making air bubbles. Components can be stored separate at −20°C in a tightly closed plastic container, sealed with parafilm. Gently thaw the separate components at room temperature before opening the vial to avoid any condensation water entering the vials. It is absolutely crucial that the epon as well as the ovens used for vacuum and polymerization are never in contact with water. Any moisture in the epon mixture will prevent appropriate polymerization.
11. Mix the components well by stirring gently with a magnetic stirrer. Avoid making air bubbles. Always make fresh, starting from thawed or from freshly made Epon A and Epon B.
12. When raising the embryos at standard water temperature of 28.5°C, the embryos hatch at approximately 48 hpf. For many reasons, this might be delayed. If they did not hatch at the time of fixation, they need to be dechorionated to make sure that the probe or antibody can reach the tissue. Dechorionation of zebrafish embryos can be done manually with forceps and scalpel

or by enzymatic pronase treatment. We prefer manual dechorionation.

13. If you do this depigmentation step in an eppendorf, make sure to make a little hole in the cap of the eppendorf as the combination of KOH and H_2O_2 produces a small amount of gas. Without a hole in the cap, the eppendorf may explode and the material will be lost. It is also possible to work in a well plate. As the embryos lose their pigments, they become very hard to see. Putting the well plate on a black background will facilitate the observation of the colorless samples.
14. To determine the concentration of proteinase K, always use the stage closest to the stage of the embryo or larva you are using. For example, larvae of 68 hpf will be treated with 2× proteinase K for 40 min.
15. There is no separate step in which endogenous peroxidase is blocked. This is because during depigmentation, the embryos are already treated with H_2O_2 and therefore endogenous peroxidase is already blocked.
16. The working concentration needs to be determined for each primary or secondary antibody separately. Always check the corresponding data sheet of the antibody, as it will contain more information about the optimal concentration of the antibody. Generally, a good starting point is 1:300.
17. Washing steps are performed on a rocking plate, medium speed. The eppendorfs are placed flat on the rocking plate so that washing can be more intense and thus reduce aspecific binding of the primary or secondary antibody.
18. The ideal concentration depends mainly on the probe but also on the tissue. Concentrations are commonly between 0.4 and 2 μg/mL hybridization solution.
19. To reduce background, an additional RNase treatment can be performed after this washing step. Wash for 30 min with RNAse A (20 μg/mL) in 2×SSC. Wash with 2×SSC for 10 min at 37°C and wash with 2×SSC for 30 min at 65°C. Then, continue the protocol with step 23.

Acknowledgments

We thank Tommy D'Heuvaert for technical assistance. This work was supported by a GOA research grant (BOF08/GOA/019) to E.S. and A.H. B.V. acknowledges a grant of the Agency for Innovation by Science and Technology (IWT).

References

1. Huysseune, A., and Thesleff, I. (2004) Continuous tooth replacement: the possible involvement of epithelial stem cells, *Bioessays 26*, 665–671.
2. Borday-Birraux, V., Van der heyden, C., Debiais-Thibaud, M., Verreijdt, L., Stock, D. W., Huysseune, A., and Sire, J. Y. (2006) Expression of Dlx genes during the development of the zebrafish pharyngeal dentition: evolutionary implications, *Evol Dev 8*, 130–141.
3. Jackman, W. R., Draper, B. W., and Stock, D. W. (2004) Fgf signaling is required for zebrafish tooth development, *Developmental Biology 274*, 139–157.
4. Jackman, W. R., Yoo, J. J., and Stock, D. W. (2010) Hedgehog signaling is required at multiple stages of zebrafish tooth development, *BMC Dev Biol 10*, 119.
5. Wise, S. B., and Stock, D. W. (2010) bmp2b and bmp4 are dispensable for zebrafish tooth development, *Dev Dyn 239*, 2534–2546.
6. Verstraeten, B., Sanders, E., van Hengel, J., and Huysseune, A. (2010) Expression pattern of E-cadherin during development of the first tooth in zebrafish (*Danio rerio*), *Journal of Applied Ichthyology 26*, 202–204.
7. Verstraeten, B., Sanders, E., van Hengel, J., and Huysseune, A. (2010) Zebrafish teeth as a model for repetitive epithelial morphogenesis: dynamics of E-cadherin expression, *BMC Dev Biol 10*, 58.
8. Yen, A. H., and Yelick, P. C. (2011) Dental tissue regeneration – a mini-review, *Gerontology 57*, 85–94.
9. Stock, D. W. (2007) Zebrafish dentition in comparative context, *J Exp Zoolog B Mol Dev Evol 308*, 523–549.
10. Raymond, P. Double-label in situ hybridization (whole mounts), Ann Arbor, MI, USA. http://www.mcdb.lsa.umich.edu/labs/praymond/dbl_label.html.
11. Luft, J. H. (1961) Improvements in epoxy resin embedding methods, *J Biophys Biochem Cytol 9*, 409–414.
12. Verreijdt, L., Debiais-Thibaud, M., Borday-Birraux, V., Van der heyden, C., Sire, J. Y., and Huysseune, A. (2006) Expression of the dlx gene family during formation of the cranial bones in the zebrafish (*Danio rerio*): Differential involvement in the visceral skeleton and braincase, *Developmental Dynamics 235*, 1371–1389.
13. Van der heyden, C., and Huysseune, A. (2000) Dynamics of tooth formation and replacement in the zebrafish (*Danio rerio*) (Teleostei, Cyprinidae), *Dev Dyn 219*, 486–496.

INDEX

Chrissa Kioussi (ed.), *Odontogenesis: Methods and Protocols*, Methods in Molecular Biology, vol. 887,
DOI 10.1007/978-1-61779-860-3, © Springer Science+Business Media, LLC 2012

MIX
Papier aus verantwortungsvollen Quellen
Paper from responsible sources
FSC® C105338

If you have any concerns about our products,
you can contact us on
ProductSafety@springernature.com

In case Publisher is established outside the EU,
the EU authorized representative is:
Springer Nature Customer Service Center GmbH
Europaplatz 3, 69115 Heidelberg, Germany

Printed by Libri Plureos GmbH
in Hamburg, Germany